AF771565

Drought - Detection and Solutions

Edited by Gabrijel Ondrasek

Published in London, United Kingdom

IntechOpen

Supporting open minds since 2005

Drought - Detection and Solutions
http://dx.doi.org/10.5772/intechopen.78865
Edited by Gabrijel Ondrasek

Contributors
Olga Panfilova, Sergey Knyazev, Olga Golyaeva, Olga Kalinina, Leonardo Ornella, Jose Crossa, Amber Hoover, Rachel M. Emerson, Allison Ray, Damon Hartley, Jason Hansen, Matilde García-Valdecasas Ojeda, Sonia Raquel Gamiz-Fortis, María Jesús Esteban-Parra, Yolanda Castro-Díez, Emilio Romero-Jiménez, Yuriy Kuleshov, Parvaze Sofi, Musharib Gull, Khalid Rehman, Ricardo Trejo-Calzada, Aurelio Pedroza-Sandoval, Jesus G. Arreola-Avila, Fabian García-González, Ivan Mustac, Vilim Filipović, Lana Filipovic, Gabrijel Ondrasek, Dragutin Petošić, Katalin Posta, Nguyen Hong Duc, Jean-Philippe Terreaux, Mabel Tidball, Vladimir Meglič, Aleš Sedlar, Marjetka Kidrič, Jelka Šuštar Vozlič, Barbara Pipan, Tanja Zadražnik

© The Editor(s) and the Author(s) 2020
The rights of the editor(s) and the author(s) have been asserted in accordance with the Copyright, Designs and Patents Act 1988. All rights to the book as a whole are reserved by INTECHOPEN LIMITED. The book as a whole (compilation) cannot be reproduced, distributed or used for commercial or non-commercial purposes without INTECHOPEN LIMITED's written permission. Enquiries concerning the use of the book should be directed to INTECHOPEN LIMITED rights and permissions department (permissions@intechopen.com).
Violations are liable to prosecution under the governing Copyright Law.

CC BY

Individual chapters of this publication are distributed under the terms of the Creative Commons Attribution 3.0 Unported License which permits commercial use, distribution and reproduction of the individual chapters, provided the original author(s) and source publication are appropriately acknowledged. If so indicated, certain images may not be included under the Creative Commons license. In such cases users will need to obtain permission from the license holder to reproduce the material. More details and guidelines concerning content reuse and adaptation can be found at http://www.intechopen.com/copyright-policy.html.

Notice
Statements and opinions expressed in the chapters are these of the individual contributors and not necessarily those of the editors or publisher. No responsibility is accepted for the accuracy of information contained in the published chapters. The publisher assumes no responsibility for any damage or injury to persons or property arising out of the use of any materials, instructions, methods or ideas contained in the book.

First published in London, United Kingdom, 2020 by IntechOpen
IntechOpen is the global imprint of INTECHOPEN LIMITED, registered in England and Wales, registration number: 11086078, 7th floor, 10 Lower Thames Street, London, EC3R 6AF, United Kingdom
Printed in Croatia

British Library Cataloguing-in-Publication Data
A catalogue record for this book is available from the British Library

Additional hard and PDF copies can be obtained from orders@intechopen.com

Drought - Detection and Solutions
Edited by Gabrijel Ondrasek
p. cm.
Print ISBN 978-1-78984-780-2
Online ISBN 978-1-78984-781-9
eBook (PDF) ISBN 978-1-83968-354-1

For EU product safety concerns:
IN TECH d.o.o., Prolaz Marije Krucifikse Kozulić 3, 51000 Rijeka, Croatia, info@intechopen.com or visit our website at intechopen.com.

We are IntechOpen,
the world's leading publisher of
Open Access books
Built by scientists, for scientists

4,500+
Open access books available

119,000+
International authors and editors

135M+
Downloads

151
Countries delivered to

Our authors are among the
Top 1%
most cited scientists

12.2%
Contributors from top 500 universities

WEB OF SCIENCE™

Selection of our books indexed in the Book Citation Index
in Web of Science™ Core Collection (BKCI)

Interested in publishing with us?
Contact book.department@intechopen.com

Numbers displayed above are based on latest data collected.
For more information visit www.intechopen.com

Meet the editor

Gabrijel Ondrasek, PhD, is employed at the University of Zagreb, Faculty of Agriculture, Croatia, as a full professor and head of the Department of Soil Amelioration. His academic and scientific opus is orientated towards sustainable management of natural resources (water, soils) in (agro)ecosystems, notably those that are exposed to disturbed water balance on the soil-plant route, excessive salinity, and metals contamination. He has edited four books and published seventeen book chapters and more than thirty scientific papers.

Contents

Preface

Annually, the global hydrological cycle circulates a large amount (~0.6 Mkm^3) of freshwater that is returned as precipitations (*blue and green water*) into the aquatic (~80%) and terrestrial (~20%) ecosystems, causing spatiotemporal distribution of readily available, fresh hydro-resources. The annual per capita availability of fresh water resources has sharply reduced during the last century, while over the same period water withdrawals and consumption has increased in the agri-/domestic/industrial triangle. In the same period availability and quality of fresh (*blue/green*) water resources have been exposed to extreme pressures (pollution, population growth, urbanization, deruralization) and non-sustainable management (overexploitation, rising of *grey waters*), which, under global climate changes, further disturbs the water cycling/balance. As a consequence, global ecosystems are frequently experiencing negative water balance (-ΔW), that is, a scenario in which certain domain water inputs (precipitations, river/groundwater inflows) are substantially lower than water outputs (evapotranspiration, water abstraction, deep percolation). Although drought and aridity are not synonymous, both share –ΔW. However, while drought assumes a relatively short-term –ΔW (usually several weeks to months), that is, a temporary, recurring reduction of water level/volume/flow in a certain domain (river, lake, catchment, aquifer), aridity implies long-term –ΔW, that is, permanent average climate condition with negative water balance. Irrespective of drought situations (hydrological, meteorological, agronomical), they jeopardize energy and food production, ecological value of certain domains, and limit *blue/green waters* for our basic needs. Thus, each particular drought type should be managed appropriately, often with costly integrative agro-hydrotechnical approaches. New and more sustainable approaches are being developed depending on environmental conditions. This book presents the most recent insights related to drought types, their detection, and their effects on food, energy, and municipal water supplies. It also examines some novel approaches to drought management, which is one of the most challenging tasks for humankind.

Gabrijel Ondrasek
University of Zagreb,
Faculty of Agriculture,
Zagreb, Croatia

Chapter 1

Satellite Data and Supervised Learning to Prevent Impact of Drought on Crop Production: Meteorological Drought

Leonardo Ornella, Gideon Kruseman and Jose Crossa

Abstract

Reiterated and extreme weather events pose challenges for the agricultural sector. The convergence of remote sensing and supervised learning (SL) can generate solutions for the problems arising from climate change. SL methods build from a training set a function that maps a set of variables to an output. This function can be used to predict new examples. Because they are nonparametric, these methods can mine large quantities of satellite data to capture the relationship between climate variables and crops, or successfully replace autoregressive integrated moving average (ARIMA) models to forecast the weather. Agricultural indices (AIs) reflecting the soil water conditions that influence crop conditions are costly to monitor in terms of time and resources. So, under certain circumstances, meteorological indices can be used as substitutes for AIs. We discuss meteorological indexes and review SL approaches that are suitable for predicting drought based on historical satellite data. We also include some illustrative case studies. Finally, we will survey rainfall products existing at the web and some alternatives to process the data: from high-performance computing systems able to process terabyte-scale datasets to open source software enabling the use of personal computers.

Keywords: remote sensing, supervised learning, meteorological index, wavelet

1. Introduction

Climate change is shifting the rainfall patterns and increasing the severity of droughts and floods around the Earth. Australia [1], Europe, and the rest of the continents have been affected by a number of major drought events [2]. In 2018, drought and heat waves reduced harvests up to 40–50% in some countries of northern and central Europe [3].

Drought is by far the Earth's most costly natural disaster and can have widespread impacts [4]. Globally, it is responsible for 22% of the economic damage caused by natural disasters and 33% of the damage in terms of the number of people affected [5]. Though average yields rose steadily between 1947 and 2008, there is no evidence that relative stress tolerance has improved [6, 7]. Therefore, until breeding programs develop adapted germplasm, drought forecasting will be

IntechOpen

important to determine when to take contingency actions to prevent drought and mitigate its risk and impacts.

The practice of drought forecasting remains challenging and is subject to great uncertainty partly due to the instability of the components of the hydrologic cycle (e.g., rainfall, soil moisture, groundwater level, etc.); temporal variability involving trends, oscillating behavior, and sudden shifts that appear in hydroclimatic records, thus posing challenges to drought prediction [8, 9].

Although agricultural indices (AIs) better reflect the soil water situation that influences crop conditions, monitoring of soil moisture is costly in terms of time and resources [10]. On the other hand, some meteorological indices (e.g., the Standard Precipitation Index) can be calculated just knowing the precipitation (pp) data, and then the expert can give a very close condition of the vegetation [11].

A variety of methods has been developed to predict drought occurrence: statistical run theory [12], Markov chain [13], loglinear [14], renewal process [15], and Poisson process [16], among others.

A valuable alternative to the aforementioned methods is machine learning (ML), a branch of artificial intelligence that studies how to extract information from big data sets with minimal human intervention. ML has been successfully tested in very different areas such as bioinformatics [17], crop protection [18], and economics [19], among others. Therefore, its potential for predicting the climate seems far from being fully exploited.

The remainder of this chapter is organized as follows: in Section 2, we introduce some representative ML methods that have been proposed for drought forecasting; in Section 3, we present the concept of meteorological drought, in particular the standardized precipitation index that is considered a primary drought indicator. Section 4 describes some forecast examples using the abovementioned methods. In Sections 5 and 6, we review satellite precipitation products and how to access and process them; and finally, in Section 7 we present the conclusions of this work.

2. Machine learning

ML is the science of algorithms and statistical models that computer systems use to progressively improve their performance of a specific task. They can be broadly categorized into supervised and unsupervised learning. In SL (classification or regression), the algorithm builds a function from a set of data relating the inputs to the outputs. In regression, the outputs are continuous, meaning they may have any value within a range (e.g., temperature and moisture), while in classification, the outputs are restricted to a limited set of values.

In unsupervised learning, the algorithm builds a mathematical model of a data set that contains inputs and no outputs. These unsupervised learning procedures are used to find structure in the data (e.g., cluster data) or reduce its dimensionality.

Examples of ML are land classification using remote sensing [20–22], amending satellite data assimilation [23], or decomposing the causes of climate change [24].

2.1 Support vector regression (SVR) and least squares support vector regression (LS-SVR)

SVR is based on the Vapnik-Chervonenkis (VC) theory [25], which characterizes the properties of learning machines that enable them to generalize the unobserved data well.

Starting with the simplest example, that is, linear regression, the objective of both SVR [26] and LS-SVR [27] is to fit a linear relation $y = \boldsymbol{w}^T\boldsymbol{x} + b$ between the x regressors and the dependent variable y in the so-called feature space. In SVR, the problem is solved by minimizing

$$\frac{1}{2}\|w\|^2 + C\sum_{i=1}^{l}\left(\xi_i + \xi_i^*\right) \tag{1}$$

under the constraints

$$\begin{gathered} y - w^T x_i - b \leq \varepsilon + \xi_i \\ y - w^T x_i - b \leq \varepsilon + \xi_i{}^* \\ \xi_i, \xi_i^* \geq 0 \end{gathered} \tag{2}$$

while for LS-SVR, the objective is to minimize

$$\|\boldsymbol{w}\|^2 + \gamma\sum_{i=1}^{n} e_i^2 \tag{3}$$

under the constraints

$$y_i - w^T x_i - b \tag{4}$$

Both methods are very similar, but in LS-SVR, the objective is to minimize the more usual sum of the squares of the errors, by replacing the ε-tube or ε-insensitive loss of SVR, that is, by ignoring all regression errors smaller than ε (**Figure 1**). Solving a nonlinear regression demands a “kernel trick” [26]. This trick uses kernel functions to transform the data of the input space into a higher dimensional feature space to make it possible to perform a linear regression. Common kernels are

$$\text{polynomial } k\left(x_i, x_j\right) = \left(x_i \cdot x_j\right)^d \tag{5}$$

$$\text{Gaussian radial basis function } k\left(x_i, x_j\right) = \exp\left(-\frac{\left\|x_i - x_j\right\|^2}{2\sigma^2}\right) \tag{6}$$

LS-SVR is an economic alternative to the original SVR model. It only relies on the cost function on a sum-of-squared-error (SSE) and equality constraints, instead of the computationally complex and time-consuming quadratic programming problem in SVR [28].

For optimal performance, parameter tuning is necessary [29]: for SVR, C and ε and the kernel-related parameters (e.g., σ^2 for the RBF kernel) and for LS-SVR, g (the regularization parameter determining the trade-off between the fitting error and the smoothness of the estimated function) and the kernel-related parameters. For further information about SVR in general, the reader should refer to [30].

2.2 Artificial neural network (ANN)

An ANN is a supervised learning model based on the operation of biological neurons. There are many architectures and training algorithms for ANN. The multilayer perceptron network (MLPN), the most common ANN architecture used for forecasting, consists of a feedforward neural network with at least three layers of neurons: an input layer, one or more hidden layers, and an output layer with a directed acyclic graph representation network (**Figure 2**). The input layer receives the data vector x, while the output layer gives the output vector y. An activation function is applied to activate the neurons in the hidden layer. For a three-layer network system, the nonlinear mapping between input x and output y is given by the equation:

$$y = \frac{1}{2}\left[\sum_{j=1}^{h}\left(w_j f_1 \sum_{i=0}^{n} w_{ji} x_i\right)\right] \tag{7}$$

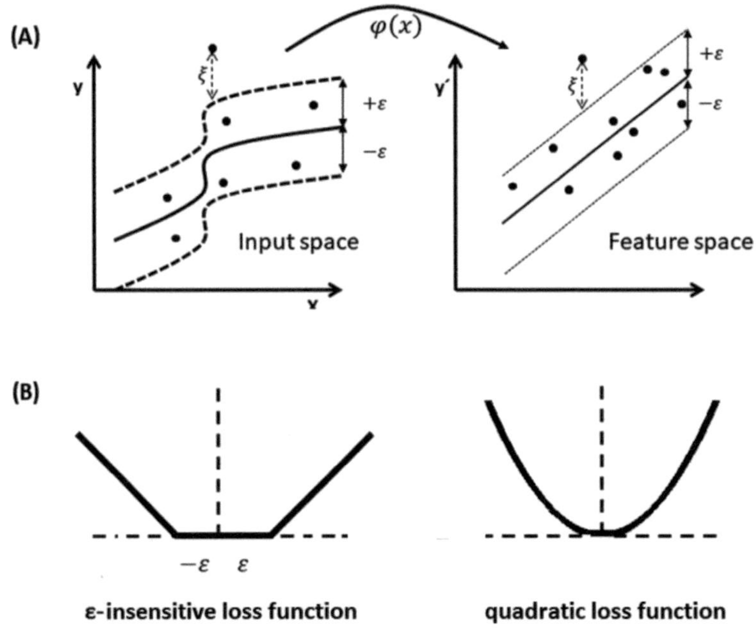

Figure 1.
(A) Kernel trick: mapping the data from the input space into a feature space. (B) Loss function used in support vector regression (ε-insensitive loss) and least squares support regression (quadratic).

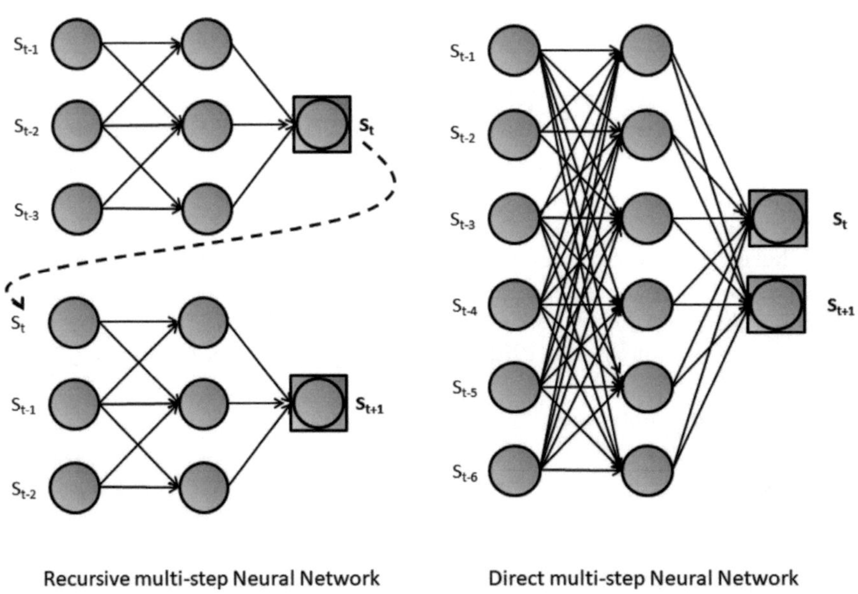

Figure 2.
Architectures of forecasting artificial neural networks. Recursive multistep neural network versus direct multistep neural network.

An ANN is usually learned by adjusting the weights and biases in order to minimize a cost function, usually MSE using the error back-propagation algorithm.

Of the activation functions, we should mention the hyperbolic tangent$f(x) = \frac{e^x - e^{-x}}{e^x + e^{-x}}$ and the sigmoidal function: $(x) = \frac{e^x}{1+e^{-x}}$.

The number of hidden neurons is no less important, since a wrong number may cause either overfitting or underfitting problems. Normally it is selected via trial and error, but this is computationally costly. Several heuristics or formulas have been proposed to avoid this cumbersome work, and success depends on the type of data, the complexity of the network architecture, etc. [31].

Last but not least, ANN forecasting models can be separated into two broad groups, namely, the recursive multistep neural network (RMSNN) and the direct multistep neural network (DMSNN) (**Figure 2**). In RMSNN, the model forecast one time-step ahead, and the network is applied recursively, using previous predictions as inputs for subsequent forecasts—that is, a forecast horizon of 3 months will have, as inputs, the outputs of forecasts with lead times of 1 and 2 months.

Similar to the RMSNN model, the DMSNN approach has a single or multiple neurons in both the input and hidden layers. However, it can have several neurons in the output layer representing multiple-month lead time forecasts. Similar to the RMSNN model, the DMSNN model is designed to forecast drought conditions using the present index value and several months of past index values as inputs.

2.3 Deep belief networks (DBN)

ANNs are suitable for complex time series forecasting but have several weaknesses: (1) selection of the initial values of the weights (normally at random) can affect the learning process, leading to slower convergence or to different forecast results for each training process and (2) the training process may get stuck at local optima, especially in networks with several hidden layers. Hinton et al. [32] proposed a probabilistic generative model with multiple hidden layers that uses layer-wise unsupervised learning to pre-train the initial weights of the network and then fine-tune the whole network using standard supervised methods such as the back-propagation algorithm.

Classically, a DBN is constructed by stacking multiple restricted Boltzmann machines (RBMs) on top of each other (**Figure 3**). The layers are trained by using the feature activations of one layer as the training data for the next layer. Better initial values of weights in all layers are obtained by greedy layer-wise unsupervised training, and the entire network is fine-tuned using an SL algorithm. Pre-training can be done with principal component analysis or nonlinear generalization [33].

An RBM [34] is a neural network model used for unsupervised learning. Typically, it consists of a single layer of hidden units (the outputs) with undirected and symmetrical connections to a layer of visible units (the data) (**Figure 3**). The configuration (bipartite graph) defines the state of each unit. Only connections between a hidden unit and a visible unit are permitted—that is, no connections between two visible units or between two hidden units are allowed. An RBM is a special type of generative energy-based model that is defined in terms of the energies of the configurations between visible and hidden units.

The standard type of RBM has binary-valued (Boolean/Bernoulli) hidden and visible units.

2.4 Bagging

Bootstrap aggregating, or bagging, is an ML ensemble meta-algorithm designed to increase the stability and accuracy of unstable procedures, for example, artificial

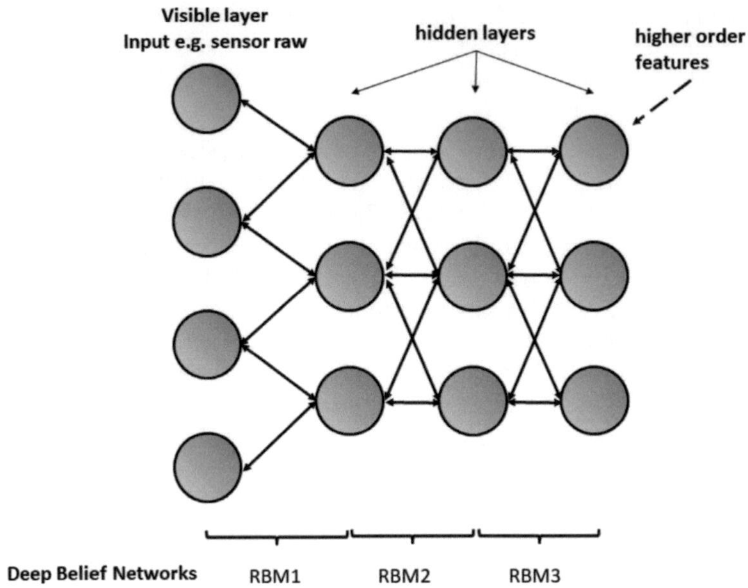

Figure 3.
Basic deep belief network (DBN) structure with three hidden layers.

neural networks or decision trees [35]. Given a standard training set T of size n, the algorithm sample is taken from T uniformly and with replacement m new training sets, T', each of size n′ (some observations may be repeated in each D). This process is known as a bootstrap sampling [36]. The basic idea is that the samples are de-correlated, and this reduces the expected error as m increases.

The *m* models are fitted using the above *m* bootstrap samples, and results of an unknown instance are obtained by averaging the output (for regression) or by voting (for classification) (**Figure 4**).

This method may slightly degrade the performance of stable algorithms (e.g., k-nearest neighbor) because smaller training sets are used to train each algorithm.

Bagging does not necessarily improve forecast accuracy in all cases. Nevertheless, this method and its derivatives tend to outperform traditional forecasting procedures [37].

2.5 Random forest regression (RFR)

A random forest (RF) [38] is a collection of *K* binary recursive partitioning trees, where each tree is grown on a subset of n instances extracted with replacement from the original training data. It is an instance of bagging where the individual learners are de-correlated trees. Each tree is grown in a top-down recursive manner, from the root node to terminal nodes or leaves (**Figure 5**). In each node, a random sample of m ($m \approx p/3$) predictors is chosen as candidates from the full set of p predictors. The data are partitioned into the two descendant branches by choosing the variable that minimized:

$$RSS = \sum_{left} (y_i - y_L{}^*)^2 - \sum_{right} (y_i - y_R{}^*)^2 \quad (8)$$

The advantage of selecting a random subset of predictors is that two trees generated on same training data will be de-correlated (independent of each other) because randomly different variables were selected at each split. Each internal (non-leaf) node is signed with a predictor determined by the *RSS* test, and each one of the two possible subsets of this variable labels the arcs connecting to the subordinate decision node. Each tree extends as much as possible until all the terminal nodes are maximally homogeneous (a minimum of five examples in each leaf is recommended).

Once the random forest is generated, the output of new data is obtained by averaging the predictions of the K trees.

The number of trees influences the error of prediction; it decreases as the number of trees (ntree) grows, but there is a threshold beyond which there is no significant gain [38, 39]. In general, ntree≈500 gives good results [40].RF can

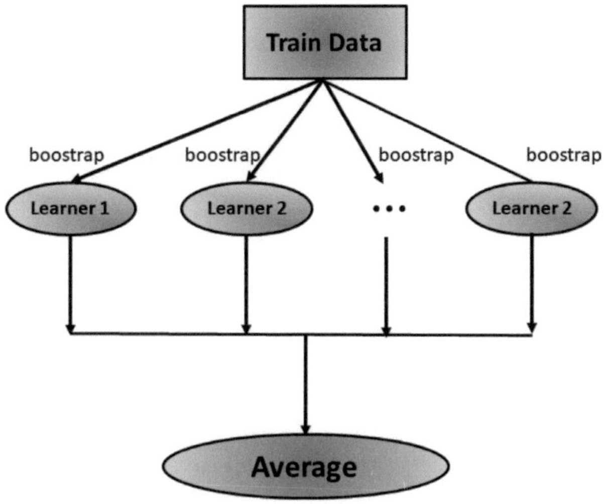

Figure 4.
Structure of bootstrap aggregating, or bagging.

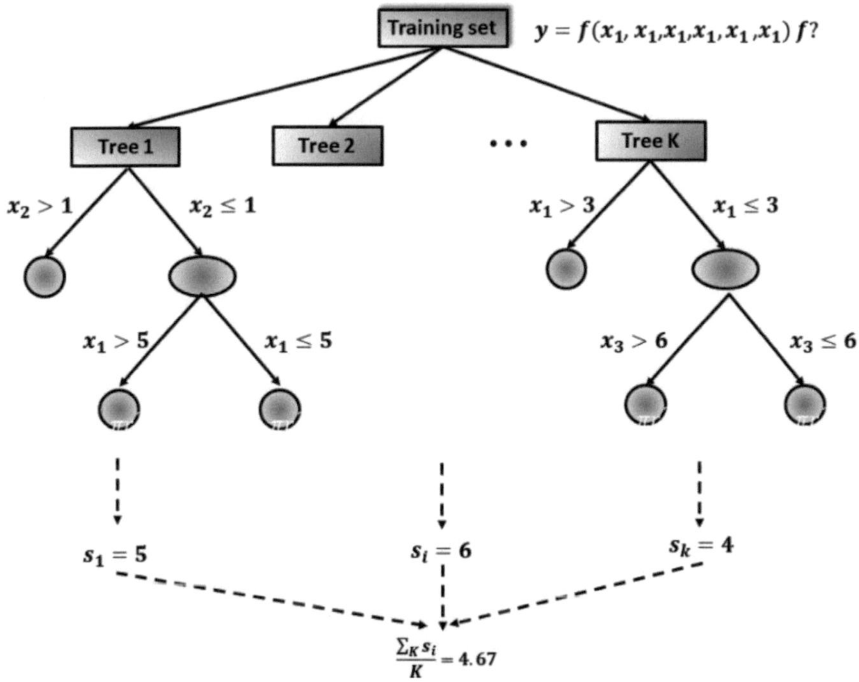

Figure 5.
Architecture of the random forest model.

successfully handle high dimensionality and multicollinearity, because it is both fast and insensitive to overfitting. It is, however, sensitive to the sampling design.

2.6 Adaptive neuro-fuzzy inference system or adaptive network-based fuzzy inference system (ANFIS)

ANFIS is a hybrid learning procedure which employs the linguistic concept of fuzzy systems (human knowledge) and the training power of the ANN to solve a regression problem [41]. All ANFIS works reported here are based on the Takagi-Sugeno fuzzy inference system [42], where the fuzzy rule applied has the form: if x is A and y is B then $z = f(x, y)$. Other fuzzy methods are Mamdani-type or Tsukamoto-type [42].

Figure 6 depicts a typical ANFIS architecture. Square nodes (adaptive nodes) have parameters, while circle nodes (fixed nodes) do not. The first and the fourth layers contain the parameters that can be modified over time. A particular learning method was required to update these parameters.

In layer 1, every node is adaptive and associated with an appropriate continuous and piecewise differentiable function such as Gaussian, generalized bell-shaped, trapezoidal-shaped, and triangular-shaped functions.

In layer 2, every node is fixed and represents the firing strength of each rule. This is calculated by the fuzzy and connective method of the "product" of the incoming signals, that is, $O_i^2 = w_i = \mu_{Ai}(x) * \mu_{Bi}(x), i\_1, 2$.

In layer 3, every node is also fixed, showing the normalized firing strength of each rule. The ith node calculates the ratio of the ith rule's firing strength to the summation of two rules' firing strengths.

In every adaptive node of layer 4 (consequent nodes) is a function indicating the contribution of the ith rule to the overall output: $O_{4,i} = \overline{w_i} f = \overline{w}_i(p_i + q_i + r_i)$, where w_i is the output of layer 3 and p_i, q_i, r_i is the parameter set. Finally, layer 5 (output node) is a single node that computes the overall output of the ANFIS as: $O_{5,1} = \sum \overline{w_i} f_i = \frac{\sum_i w_i f_i}{\sum_i w_i}$.

One of the most important steps in developing a satisfactory forecasting model is the selection of the input variables. These variables determine the structure of the forecasting model and affect the weighted coefficients and the results of the model

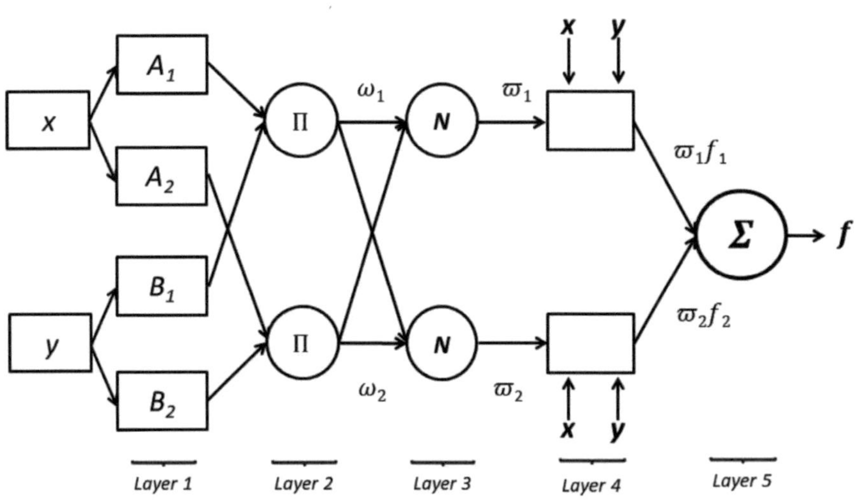

Figure 6.
Architecture of an adaptive network-based fuzzy inference system (ANFIS).

function in layer 2. As the number of parameters increases with the fuzzy rule increment, the model structure becomes more complicated. A very good description of ANFIS is presented in [43, 44].

2.7 Boosting

Boosting attempts to increase the performance of a given learning algorithm by iteratively adjusting the weight of an observation based on the last training/testing process. In other words, the meta-algorithm produces a sequence of models by adaptive reweighting of the training set [45].

AdaBoost, the first boosting algorithm, is definitely beaten by noisy data; its performance is highly affected by outliers, as the algorithm tries to fit every point perfectly. Friedman [46] extended the concept to present gradient boosting, which constructs additive regression models by sequentially fitting a simple parameterized function (base learner) to current "pseudo"-residuals by least squares at each iteration. The pseudo-residuals are the gradient of the loss function being minimized with respect to the model values at each training data point evaluated in the current step. This reduces the loss of the loss function. We iteratively added each model and computed the loss. The loss represents the difference between the actual value and the predicted value (the error residual), and using this loss value, the predictions are updated to minimize these residuals.

A regularization method that penalizes various parts of the boosting algorithm is necessary to avoid overfitting. This generally improves the performance of the algorithm by reducing overfitting.

2.8 Hybrid models

The time series that characterize the evolution of meteorological events (drought, precipitation) in the temporal domain have localized high- and low-frequency components with dynamic nonlinearity and non-stationary features. MM models have not always proven to be good at capturing the behavior of the time series. Hybrid models can perform superbly when forecasting hydrological and climatological time series. Different combination techniques have been proposed in order to overcome the deficiencies of single models and improve forecasting performance [47]. Many combined models have been introduced in the literature, for example, ANN-ARIMA [48], SVR-ARIMA [49], etc.

Here we will only focus on WT-ML hybrids, where ML is a machine learning method (e.g., ANN or SVR) and WT is a discrete wavelet transform [50].

2.8.1 Wavelet transform (WT)

WT is a time-dependent spectral analysis that decomposes time series in the time-frequency space and provides a timescale illustration of processes and their relationships. In this method, the data series are broken down by transforming them into "wavelets," which are scaled and shifted versions of a mother wavelet [50]. This allows the use of long time intervals for low-frequency information and shorter intervals for high-frequency information and can reveal aspects of data such as tendencies, breakdown points, and discontinuities that other signal analysis techniques might miss, for example, Fourier transform.

There are two main alternatives for WT: discrete wavelet transform (DWT) and continuous wavelet transform (CWT). For DWT, the WT is applied using a discrete set of the wavelet scaling and shifting, whereas in the case of CWT, this scaling and shifting is continuous—that is, CWT is computationally expensive

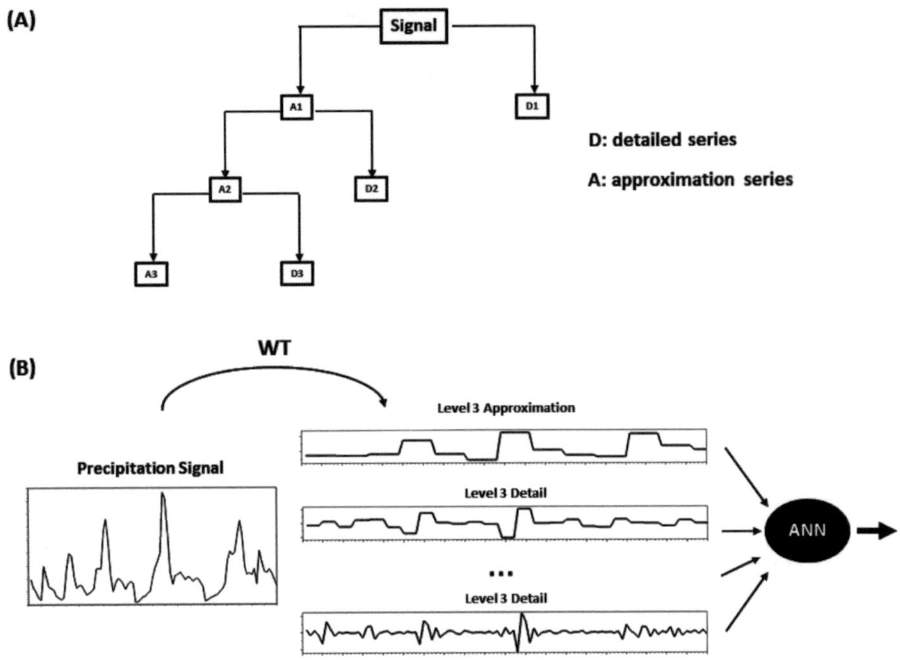

Figure 7.
Time series wavelet-ANN conjunction model. (A) Three-level wavelet decomposition tree (DWT). (B) Example of the decomposition of a precipitation signal.

and most researchers use DWT. For more information about CWT, the reader should refer to [51].

DWT operates two sets of functions (scaling and wavelets) viewed as high-pass (HPF) and low-pass (LPF) filters. The signal is convolved with the pair of HPF and LPF followed by subband downsampling producing two components. The first component, which is obtained by passing the signal through the low-pass filter, is called an approximation component (or series), and the other component (fast events) is called a detailed component (**Figure** 7). This process is iterated n times with successive approximation series being decomposed in turn, so that the original time series is broken down into the minimum number of components needed to reflect the time series according to the mother wavelet.

The filterbank implementation of wavelets can be interpreted as computing the wavelet coefficients of a discrete set of child wavelets for a given mother. This mother wavelet function was defined at scale a and location b as

$$\psi_{a,b}(t) = \frac{1}{\sqrt{a}}\psi\left(\frac{t-b}{a}\right) \tag{9}$$

$\psi_{0,0}(t)$ is a mother wavelet prototype and a, b are scaling and shifting parameters, respectively.

Several wavelet families have proven useful for forecasting various hydrological time series. As an example, we can mention Haar, which is also known as daubechies1 or db1 [50]. It is defined as

$$\begin{cases} 1 \; if \; 0 < t < 0.5 \\ -1 \; if \; 0.5 < t < 1 \\ 0 \quad otherwise \end{cases} \tag{10}$$

A full description of DWT can be found in [50, 52].

3. Meteorological indices

Drought can be defined as a period of unusually arid conditions (usually due to rainfall deficiency) that have lasted long enough to cause non-balance in a region's hydrological situation. Based on its intensity and persistence, drought can be classified into four categories [53]: (1) meteorological drought, which occurs when precipitation is less than usual, is characterized by changes in weather patterns; (2) agricultural (vegetation) drought refers to water deficits in plants; it occurs after meteorological drought and before hydrological drought; (3) hydrological drought ensues when the level of surface water and the groundwater table are less than the long-term average; and finally, (4) socioeconomic drought materializes when water resources required for industrial, agricultural, and household consumption are less than required and thus cause socioeconomic anomalies.

A drought index is an indicator or measure derived from a series of observations that reveals some of the cumulative effects of a prolonged and abnormal water deficit. It integrates pertinent meteorological and/or hydrological parameters (accumulated precipitation, temperature, and evapotranspiration) into a single numerical value or formula and gives a comprehensive picture of the situation [53]. Such an index is more readily usable and comprehensible than the raw data and, if presented as a numerical value, makes it easier for planners and policymakers to make decisions. Authorities and public and private committees evaluate the impact of drought using these indices and take measures to prevent its effects [54].

More than 100 drought indices have so far been proposed, and each one has been formulated for a specific condition [55]. The reclamation drought index (RDI), for example, was developed in the USA to activate drought emergency relief funds associated with public lands affected by drought; the crop moisture index (CMI) was designed to show the effects of water conditions on growing crops in the short term and is not a good instrument for displaying long-term conditions. Here we will only describe the standardized precipitation index, which those indices used in case studies.

3.1 Standardized precipitation index (SPI)

Most of the forecasting works reviewed here are based on SPI [56]. It is perhaps the most popular index for forecasting meteorological drought and has been recommended by the World Meteorological Organization [57]. It can be defined as the number of standard deviations that the observed cumulative rainfall at a given time scale (1,3,6 month) would deviate from the long-term mean for that same time scale over the entire length of the record (z-score).

More specifically, SPI is calculated by building a frequency distribution from historical precipitation data (at least 30 years) at a specific location for the precipitation accumulated during a specified period, for example, 1 month (SPI1), 3 months, (SPI3), 24 months (SPI24), and so on. A theoretical probability density function (usually the gamma distribution) is fitted to the empirical distribution for the selected time scale.

SPI1 to SPI6 are considered indices for short-term or seasonal variation (soil moisture), whereas SPI12 is considered a long-term drought index (groundwater and reservoir storage).

The "drought" part of the SPI range is arbitrarily split into "near normal" ($0.99 > \text{SPI} > -0.99$), "moderately dry" ($-1.0 > \text{SPI} > -1.49$), "severely dry"

(−1.5 > SPI > −1.99), and "extremely dry" (SPI < −2.0) conditions [56]. A drought event starts when SPI becomes negative and ends when it becomes positive again.

SPI is easy to calculate (using precipitation only) and can characterize drought or abnormal wetness on different time scales. Its standardization ensures independence from geographical position, and it is thus more comparable across regions with different climates. The index can be computed using several packages of the R project [58], for example, the SPEI package [59] or the SPI package [60]. Limitations of SPI include the following: (1) it does not account for evapotranspiration; (2) it is sensitive to the quantity and reliability of the data used to fit the distribution; and (3) it does not consider the intensity of precipitation and its potential impacts on runoff, streamflow, and water availability within the system. A more detailed explanation of how SPI is calculated can be found at [43].

3.2 Other indices

Other indices including only precipitation data are EDI [61], SIAP [62], deciles index (DI), percent of normal (PN), standard precipitation index (SPI), China-Z index (CZI), modified CZI (MCZI), and z-score [55].

4. Forecasting meteorological drought

Forecasting meteorological drought using historical data is not a trivial task. The time series that characterize the evolution of meteorological events (drought, precipitation) in the temporal domain have localized high- and low-frequency components with dynamic nonlinearity and non-stationary features. Several statistical indicators have been proposed to evaluate the success of prediction. Most of these metrics are not independent; for example, MSE can be decomposed in many ways to link it with the bias and the correlation coefficient [63]. A standard practice of model corroboration is to compute a common set of performance metrics, typically more than three. Most important is that at least three critical components, that is, one dimensionless statistic, one absolute error index statistic, and one graphical technique, should be represented in the corroboration [64].

Regarding the dimensionless statistic, we must mention:

- Pearson's correlation coefficient (R) is used to evaluate how well the estimates correspond to the observed values. Due to the standardization of many indices, the robustness of R can be limited [64].

$$R = \frac{\sum_{i=1}^{n}(p_i - \overline{p})(o_i - \overline{o})}{\sqrt{\sum_{i=1}^{n}(p_i - \overline{p})^2}\sqrt{\sum_{i=1}^{n}(o_i - \overline{o})^2}} \tag{11}$$

- Coefficient of determination (R2) measures the degree of association among the observed (o_i) and predicted values (p_i).

$$\mathrm{R2} = \frac{\sum_{i=1}^{n}(o_i - p_i)^2}{\sum_{i=1}^{n}(o_i - \overline{o})^2} \tag{12}$$

- Nash-Sutcliffe efficiency (NSE) or MSE skill [65].

$$NSE = 1 - \frac{\sum_{i=1}^{n} (p_i - o_i)^2}{\sum_{i=1}^{n} (o_i - \bar{o})^2} \tag{13}$$

- Willmott's index (WI) represents the ratio of the mean square error and the potential error [66].

$$WI = 1 - \left[\frac{\sum_{i=1}^{n} (p_i - o_i)^2}{(|p_i - o_i| + |p_i - o_i|)^2} \right] \tag{14}$$

Among one absolute error index statistic most used are

- Mean Error(MSE) estimates the average estimate error.

$$MSE = \frac{1}{n} \sum_{i=1}^{n} (p_i - o_i)^2 \tag{15}$$

- Mean absolute error (MAE).

$$\text{MAE} = \frac{1}{N} \sum_{i=1}^{n} |p_i - o_i| \tag{16}$$

In all the formulas presented above, o_i, p_i represent the observed and estimated values, n is the number of records, and $\bar{o}, \bar{p}$ indicate the means of the observed and predicted values, respectively.

Here we included R and R2, two standard regression criteria, in the group of dimensionless statistics.

Finally, we present just one example of the graphical technique, mainly to show how a training and evaluation process is executed with a ML algorithm (**Figure 8**).

4.1 Case studies

Some inconsistencies in the observations and the duration of satellite records introduce difficulties and uncertainties when applying forecast methods. At least 30 years of data record are required to SPI forecast; therefore, some of the examples we present here are based exclusively on ground gauge data. This situation is very close to reverting since satellite observations are reaching the minimum number of years required and the data are calibrated with ground observations (**Table 1**).

Shirmohammadi et al. [67] evaluated the performance of two ANN architectures (feedforward neural network and Elman or recurrent neural network), different kinds of ANFIS (four different membership functions: Gaussian, bell-shaped, triangular, and Piduetoits shape), WT-ANFIS, and WT-ANN. The wavelets families used here were db4, bior1.1, bior1.5, rboi1.1, rboi1.5, coif2, and coif4.

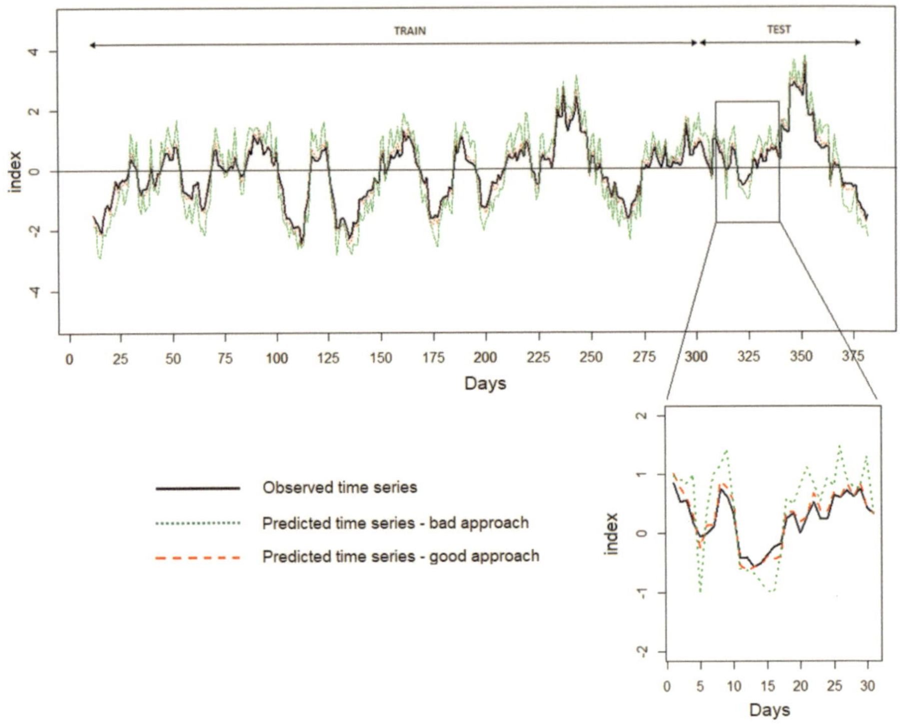

Figure 8.
Generic example of time series forecasting using two different ML methods. The green dotted line indicates a "bad" forecast method. The red dashed line indicates an appropriate method for the data, that is, the curve is closer to the observed time series. Both methods were trained using 80% of the data and tested on the remaining 20%.

Training data came from 1952 to 1992 rain records from East Azerbaijan province (Iran). More than 1000 model structures were tested to predict SPI6 for 1, 2, and 3 months' lead-time over the test period covering from 1992 to 2011. R2, NSE, and RMSE evaluated the performance of the models.

ANFIS models provided more accurate predictions than ANN models, and the inclusion of WT could improve meteorological drought modeling: WT-ANFIS (best RMSE = 0.097), WT-ANN (best RMSE = 0.227), ANFIS (best RMSE = 0.089), and ANN (best RMSE = 1.81).

Belayneh et al. [68] used precipitation records (1970 to 2005) to generate SPI3 and SPI6 time series from 12 stations in the Awash River Basin of Ethiopia (that is, 12 x 2 independent time series). The forecast was performed with ANN (RMSNN trained with the Levenberg-Marquardt back propagation), SVR, and the coupled models: WA-ANN and WA-SVR. About 80% of the data was used for training, 10% for validation, and 10% for testing, and ARIMA forecasting was used as a benchmark [69]. Regarding wavelet decomposition, each time series was decomposed between one and nine levels, and the appropriate level was selected by comparing results among all decomposition levels. The results of all the methods were compared by RMSE, MAE, and *R*2. Overall, the WA-ANN and WA-SVR models were effective in forecasting SPI3 although most WA-ANN models had more accurate estimates (1- or 3-month lead). The WA-ANN model seemed to be more effective in anticipating extreme SPI values (severe drought or heavy precipitation), whereas WA-SVR closely reflected the observed SPI trends but underestimated the extreme events.

Model	SPI3	SPI12	SPI24
SVR	0.54	0.84	0.89
BSVR	0.47	0.86	0.91
BS-SVR	0.62	0.93	0.92
ANN	0.64	0.89	0.93
BANN	0.55	0.87	0.92
BS-ANN	0.67	0.95	0.98
WBANN	0.64	0.87	0.93
WBS-ANN	0.69	0.90	0.95
WBSVR	0.57	0.85	0.90
WBS-SVR	0.67	0.95	0.94

Abbreviations: SVR, support vector regression; BSVR, bootstrap SVR ensemble; BS-SVR, boosting-SVR; ANN, artificial neural networks; BANN, bootstrap ANN ensemble; BS-ANN, boosting-ANN; WBANN, wavelet coupled bootstrap ANN ensemble; WBS-ANN, wavelet boosting-ANN; WBSVR, wavelet coupled bootstrap SVR ensemble; WBS-SVR, wavelet boosting-SVR.

Table 1.
Coefficient of determination (R2) of 10 ML methods to predict 3, 12, and 24 months SPI. Extracted from [71].

For SPI3 (1 month lead time) forecast, the best results in terms of RMSE (0.407) and MAE (0.391) were obtained by the WA-ANN model at the Ziquala station, whereas in terms of *R*2 (0.881), the Ginchi station had the best WA-ANN model.

When the lead-time was raised to 3 months, WA-ANN remained the best model. One station (Bantu Liben) had the model with the lowest RMSE and MAE values (0.510 and 0.4941), whereas a second station (Sebeta) had the best results in terms of *R*2 (0.7304).

Regarding SPI6 forecasts, the WA-ANN and WA-SVR models provided the best SPI6 forecasts. Neither method was meaningfully better than the other. The predictions for SPI6 were significantly better than SPI3 predictions according to three performance measures. As the forecast lead time increased, the forecast accuracy of all the models declined. This drop was most evident in the ARIMA, ANN, and SVR models.

These results were similar to [70]. Authors used precipitation records (1970–2005) from 20 stations in the same basin of Ethiopia (three different sub-basins) to generate SPI3 and SPI12 series. ANN, SVR, and WA-ANN were evaluated for 1 and 6 months lead time prediction. The comparison was made using RMSE, MAE, and R2. Forecasting of SPI 12, for all the models, had better performance results than predicting SPI 3, regardless of the lead time (best R2 = 0.953, WA-ANN). The performance of all the models declined when the lead time increased.

Belayneh et al. [71] modeled ANN and SVR as in [68] to forecast SPI 3, SPI12, and SPI24, but they included bootstrap (BANN and BSVR), boosting (BS-ANN, BS-SVR), wavelet coupled bootstrap ensemble (WBANN and WBSVR), and wavelet coupled boosting (WBS-ANN and WBS-SVR) in the analysis.

In general, the performances of SVR and ANN were comparable, although ANN performance was slightly higher. The inclusion of wavelets improved both techniques (wavelet decomposition denoises the time series). All models were more effective at forecasting SPI12 and SPI24 than SPI3 (**Table 1**). All boosting ensemble models were developed in MATLAB ("fitensemble" function).

The WBS-ANN and WBS-SVR models provided better prediction results than all the other types of models evaluated.

Ali, Deo, et al. [43] evaluated the performance of three models (ANFIS, M5Tree, and MPMR) to forecast SPI3, SPI6, and SPI12 calculated from a 35-year rainfall data set (1981–2015) from three (3) stations in Pakistan. SPI data were partitioned into 70% (training) and 30% (testing) periods. M5Tree is a kind of decision tree with linear regression functions on the leaves [72], whereas MPMR stands for minimax probability machine regression [73] and was also applied to benchmark the ensemble-ANFIS model. Regarding SPI3 forecast, ANFIS (R = 0.889 to 0.946) outperformed MPMR (R = 0.843 to 0.935) and M5Tree (R = 0.831 to 0.916). Similarly, SPI6 ANFIS (R = 0.968 to 0.974) outperformed M5Tree (R = 0.950 to 0.967) and MPMR (R = 0.952 to 0.970). For SPI12, ANFIS (R = 0.987 to 0.993) overcome M5Tree (R = 0.950 to 0.967) and MPMR (R = 0.984 to 0.986). The other statistics (e.g., RMSE, WI) corroborated the superior performance of ANFIS. Just as important, the ensemble-ANFIS model achieved the highest accuracy at the three stations when predicting moderate, severe, and extreme droughts.

Khosravi et al. [74] used rainfall data from the Tropical Rainfall Measuring Mission (TRMM) during 2000–2014 in the eastern district of Isfahan to generate 12-month SPI. The first 85% of the data was used to train a single-hidden layer feedforward ANN, an SVR with RBF kernel, an LS-SVR with RBF kernel, and an ANFIS method. Optimum values of SVR and LS-SVR were obtained by a grid search within the range of [10^{-3}, 10^{+3}] and [2^{-3}, 2^{+3}] for C and γ (SVR) or (10, 100 and 1000) for g and (1, 0.5 and 1) for γ (LS-SVR).

For SPI12, SVR achieved the highest accuracy (RMSE = 0.21), followed by LS-SVR (RMSE = 0.38), ANN (RMSE = 1.24), and ANFIS (RMSE = 1.36). The best ANN model consisted of three layers (input, hidden, and output) with 30, 8, and 1 neuron, respectively.

Chen et al. [75] evaluated RF and ARIMA to forecast SPI3 (short-term drought) with a 1-month lead time and SPI12 (long-term drought). Both models were developed based on data from 1966 to 1995 (four stations in China), and predictions (1 month or 6 months ahead) were made from 1996 to 2004. Overall, RF performed consistently better than ARIMA. Results also suggested that RF is more robust in predicting dry events. Finally, ARIMA lost the capacity to predict SPI12, whereas the accuracy of RF was less affected by the longer lead time.

Agana and Homaifar [76] developed a hybrid model using a denoised empirical mode decomposition [77] and DBN. The proposed method was applied to predict a standardized streamflow index (SSI) across the Colorado River basin (ten stations). The new model was compared with MLP and SVR in predicting SSI12 (1, 6, and 12 months lead time). DBN, SVR, and their hybrid versions displayed rather similar prediction errors. However, DBN and EMD-DBN outperformed all other models for two-step predictions at almost all stations. As in wavelets, the empirical mode decomposition significantly improves the quality of prediction.

Finally, we want to mention two examples where ML was directly applied to rainfall prediction.

El Shafie et al. [77] evaluated a radial basis function neural network (RBFNN) to forecast rainfall in Alexandria City, Egypt. The model was trained using rainfall data from 1960 to 2001 (four stations) and tested with data from 2002 to 2009 to predict yearly and monthly (January and December) precipitation. Regarding yearly model efficiency, R2 = 0.94 for RBFNN, whereas the control (a multiple linear regression MR model) only reached R2 = 0.21. Regarding monthly precipitation, RBFNN was very successful (R2 = 0.899 for January and R2 = 0.997 for December) as compared to the control (R2 = 0.997 and 0.34, respectively).

Sumi et al. [78] compared ANNs, multivariate adaptive regression splines or MARS [79], k-nearest neighbor [80], and SVR with RBF kernel to predict daily and monthly rainfall in Fukuoka, Japan. A preprocessed training set (1975–2004) was used to train the algorithms with extensive parameter optimization, whereas the test set covered from 2005 to 2009. For monthly rainfall, SVR produced the most accurate forecast (lowest RMSE) and the best rainfall mapping (R2 = 0.93), whereas for the daily rainfall series, the MARS method produced the best R2 value (0.99). All the metrics were calculated based on single-step ahead forecasting.

5. Satellite precipitation products (SPPs)

There is no satellite that can reliably quantify rainfall under all circumstances. However, ground observations, although reliable and with long-term records, do not provide a consistent spatial representation of precipitation, particularly on certain world regions. Therefore, satellite data become necessary, as they provide more homogeneous data quality compared to ground observations [81, 82]. To our knowledge, merged satellite-gauge products are becoming indispensable.

Precipitation data sets may be classified into one of four categories: gauge data sets (e.g., CRU TS [83], APHRODITE [84]), satellite-exclusive (e.g., CHOMPS [85]), merged satellite-gauge products (e.g., GPCP [86], TRMM3B42), and reanalysis (e.g., NCEP1/NCEP2 [87], ERA-Interim [88]). Reanalysis implies integrating irregular observations with models encompassing physical and dynamic processes in order to generate an estimate of the state of the system across a uniform grid and with temporal continuity [89].

Many studies show that satellite precipitation algorithms show different biases, detection probabilities, and missing rainfall ratios in summer and winter. Sources of error include the satellite sensor itself, the retrieval error [90], and spatial and temporal sampling [91, 92].

Algorithms that estimate rainfall from satellite observations are based on either thermal infrared (TIR) bands (inferring cloud-top temperature), passive microwave sensors (PMW), or active microwave sensors (AMW). The TIR-based approach takes into account cold cloud duration or CCD, that is, the time that a cloud has a temperature below the threshold at a given pixel [93]. The PMW-based approach takes advantage of the fact that microwaves can penetrate clouds to explore their internal properties through the interaction of raindrops [94]. AMW is what usually known as precipitation radar [95].

There is a plethora of validation studies of satellite-based rainfall estimates (SREs). Normally, these SREs are compared against ground rainfall estimates [91, 96].

Sun et al. [97] reviewed 30 currently available global precipitation (gauge-based, satellite-related, or reanalysis) data sets. The degree of variability of the precipitation estimates varies by region. Large differences in annual and seasonal estimates were found in tropical oceans, complex mountain areas, Northern Africa, and some high-latitude regions. Systematic errors are the main sources of errors over large parts of Africa, northern South America, and Greenland. Random errors are the dominant kinds of error in large regions of global land, especially at high latitudes. Regarding satellite assessments, PERSIANN-CCS has larger systematic errors than CMORPH, TRMM 3B42, and PERSIANN-CDR. The spatial distribution of systematic errors is similar for all reanalysis products [97].

Table 2 presents a comparison of several representative satellite rainfall products. More information regarding these and other products can be found in [97, 98].

Abbreviations: (IR) infrared satellite imagery, (MW) microwave estimates, (GG) ground gauges, (AMSU) Advanced Microwave Sounding Unit, (AMSU-B) Advanced Microwave Sounding Unit-B, (SSM/I) Special Sensor Microwave/Imager; (AMSR-E) Advanced Microwave Scanning Radiometer for the Earth Observing

Product	Spatial coverage	Temporal coverage	Inputs	Access
CHIRPSv2.0 Funk et al. [99]	0.05° × 0.05° 50°S–50°N	Daily, pentadal and monthly 1981 to near present	IR + GG	http://chg.geog.ucsb.edu/data/
PERSIANN Nguyen et al. [100]	0.25° × 0.25° 60°S–60°N	1,3,6 hours March 2000 to present	IR (GOES-8, GOES-10, GMS-5, Metsat-6, and Metsat-7) corrected with MW(DMSP 7, 8, and 9 and NOAA-15, 16, and 17)	https://chrsdata.eng.uci.edu/
PERSIANN-CDR Ashouri et al. [101]	0.25° × 0.25° 60°S–60°N	Daily, monthly 1983 to 2017	IR (GRIDSAT-B1) +GPCP correction	http://chrsdata.eng.uci.edu/
CMORPH Joyce et al. [102]	0.25° × 0.25° 60°S–60°N	30 min 2002 to present	SSM/I (DMSP 13, 14, and 15) AMSU-B (NOAA-15, 16, 17, and 18) AMSU-E (Aqua), TMI (TRMM) Geostationary satellite IR	http://www.cpc.ncep.noaa.gov/products/janowiak/cmorph_description.html
RFE2.0 Xie and Arkin [103]	0.1° × 0.1° 40°N–40°S 20°W–55°E	daily 2001 to present	MW (SSM/I, AMSU-B) IR (GTS)	https://iridl.ldeo.columbia.edu/SOURCES/.NOAA/.NCEP/.CPC/.FEWS/
TRMM3B42 Huffman et al. [104]	0.25° × 0.25° 50°N–50°S	3 hourly/ daily 1998 to 2015.	MW (TRMM, SSM/I, AMSR, AMSU), IR	https://pmm.nasa.gov/data-access/downloads/trmm

Table 2.
Representative satellite rainfall products.

System, (MHS) Microwave Humidity Sounder, (GPCP) Global Precipitation Climatology Centre, (GOES) Geostationary Operational Environmental Satellite, (Metsat) meteorological satellite, (NOAA) NASA-provided TIROS series of weather forecasting satellite run by the National Oceanic and Atmospheric Administration, (DMSP) Defense Meteorological Satellite Program, (GRIDSAT-B1) geostationary IR channel brightness temperature.

6. Accessing and processing the data

The capacity to acquire information from remote sensing data has been improved to an unprecedented level, accumulating overwhelming amounts of information. For example, the Google Earth Engine (GEE) [105] is updated at a rate of nearly 6000 scenes per day from active missions (a typical image 10 km by 10 km requires 50–200 million bytes of memory). Such a large amount of data requires not only vast amounts of memory data but also higher-level services with high-performance computing systems [106]. Successful experiences have already been recorded [97], but the GEE is worth mentioning [105]. GGE stores a multi-petabyte catalog of satellite imagery and geospatial data sets collected from different resources and provides high-performance computing systems that can be accessed and controlled through an Internet-accessible application programming interface (API) and an associated web-based interactive development environment (IDE). It also possess a library with more than 800 functions, ranging from simple mathematical functions to powerful geostatistical, machine learning, and image processing operations [105].

In many situations, intense computing resources are required for image processing operations [105], but more friendly solutions can be suggested to those who do have not the necessary skills.

As mentioned before, many satellite precipitation products are freely available (**Table 1**). Most of them are in network Common Data Form (netCDF) format [95]. R users can access this format using the "ncdf4" [96] or "raster" [97] packages. These data were already processed and can be used to forecast and perform complementary analyses [98]. We have already mentioned the SPEI [59] or the SPI [60] packages used to generate, for example, the SPI index.

Regarding the ML methods discussed here, almost all of them are available in packages deposited at the CRAN or CRAN-like repositories, for example, "Random Forest" package [43], "rminer" [99] that implements ANN, SVR and boosting [99], etc. A full list of packages implementing ML algorithms is available at https://cran.r-project.org/web/views/MachineLearning.html.

Finally, also available at the repositories are plenty of packages that are really helpful for visualizing and interpreting the results [107, 108].

7. Conclusion

Climate change is shifting global rainfall patterns and will increase the intensity and duration of drought around the world; this produces the need to take contingency actions to prevent the impact of famine. ML models, an evolving research area, are a valuable complement to methods previously proposed for forecasting drought. Results obtained so far for predicting meteorological indices are very satisfactory, especially with hybrid models such as WT-ANN or WT-SVR.

Most of the work that we reported here is based on the standardized precipitation index or SPI, which is a reliable measure of drought used in more than 60 countries. The leading month or the number of months over which SPI is calculated significantly influences the prediction values.

Unfortunately, many of the examples were based on ground gauge data. The brevity (and noise) of the records obstructs the use of many satellite products. However, as time progresses and data retrieval improves, satellite products will be long and accurate enough to generate reliable results.

The exponential growth of public and free satellite imagery sources and of open-source software, as well as cheaper access to cloud-based technology, will provide powerful forecasting tools to a greater number of researchers, allowing them to forecast drought before it occurs.

Conflict of interest

The authors declare that there is no conflict of interest regarding the publication of this article.

Author details

Leonardo Ornella, Gideon Kruseman and Jose Crossa*
International Maize and Wheat Improvement Center (CIMMYT), Apdo, Mexico City, Mexico

*Address all correspondence to: j.crossa@cgiar.org

IntechOpen

© 2019 The Author(s). Licensee IntechOpen. This chapter is distributed under the terms of the Creative Commons Attribution License (http://creativecommons.org/licenses/by/3.0), which permits unrestricted use, distribution, and reproduction in any medium, provided the original work is properly cited.

References

[1] van Dijk AI, Beck HE, Crosbie RS, de Jeu RAM, Liu YY, Podger GM, et al. The millennium drought in Southeast Australia (2001–2009): Natural and human causes and implications for water resources, ecosystems, economy, and society. Water Resources Research. 2013;**49**:1040-1057. DOI: 10.1002/wrcr.20123

[2] Keshavarz M, Karami E, Vanclay F. The social experience of drought in rural Iran. Land Use Policy. 2013;**30**:120-129. DOI: 10.1016/j.landusepol.2012.03.003

[3] Harris C. Heat, Hardship and Horrible Harvests: Europe's Drought Explained. Euronews [Internet]. 2018. Available from: https://www.euronews.com/2018/08/10/explained-europe-s-devastating-drought-and-the-countries-worst-hit

[4] Hao Z, AghaKouchak A, Nakhjiri N, Farahmand A. Global integrated drought monitoring and prediction system. Scientific Data. 2014;**1**:140001. DOI: 10.1038/sdata.2014.1

[5] Wilhite DA, Svoboda MD, Hayes MJ. Understanding the complex impacts of drought: A key to enhancing drought mitigation and preparedness. Water Resources Management. 2007;**21**: 763-774. DOI: 10.1007/s11269-006-9076-5

[6] Leng G, Huang M. Crop yield response to climate change varies with crop spatial distribution pattern. Scientific Reports. 2017;7:1463-1463. DOI: 10.1038/s41598-017-01599-2

[7] Gobin A. Modelling climate impacts on crop yields in Belgium. Climate research (Open Access for articles 4 years old and older). 2010;**44**:55-68. DOI: 10.3354/cr00925

[8] Milly PCD, Betancourt J, Falkenmark M, Hirsch RM, Kundzewicz ZW, Lettenmaier DP, et al. Stationarity is dead: Whither water management? Science. 2008;**319**:573-574. DOI: 10.1126/science.1151915 %JScience

[9] Mishra AK, Singh VP. Drought modeling – A review. Journal of Hydrology. 2011;**403**:157-175. DOI: 10.1016/j.jhydrol.2011.03.049

[10] Vereecken H, Huisman JA, Pachepsky Y, Montzka C, van der Kruk J, Bogena H, et al. On the spatio-temporal dynamics of soil moisture at the field scale. Journal of Hydrology. 2014;**516**:76-96. DOI: 10.1016/j.jhydrol.2013.11.061

[11] Maity R, Suman M, Verma N. Drought prediction using a wavelet based approach to model the temporal consequences of different types of droughts. Journal of Hydrology. 2016; **539**:417-428

[12] Moyé LA, Kapadia AS, Cech IM, Hardy RJ. The theory of runs with applications to drought prediction. Journal of Hydrology. 1988;**103**:127-137. DOI: 10.1016/0022-1694(88)90010-8

[13] Paulo AA, Pereira LS. Prediction of SPI drought class transitions using Markov chains. Water Resources Management. 2007;**21**:1813-1827. DOI: 10.1007/s11269-006-9129-9

[14] Moreira EE, Coelho CA, Paulo AA, Pereira LS, Mexia JT. SPI-based drought category prediction using loglinear models. Journal of Hydrology. 2008;**354**: 116-130. DOI: 10.1016/j.jhydrol.2008.03.002

[15] Loaiciga H. On the probability of droughts: The compound renewal model. Water Resources Research. 2005;**41**. DOI: 10.1029/2004WR003075

[16] Dzupire NC, Ngare P, Odongo L. A Poisson-gamma model for zero inflated

rainfall data. Journal of Probability and Statistics. 2018;**2018**:1-12. DOI: 10.1155/2018/1012647

[17] Macesic N, Polubriaginof F, Tatonetti NP. Machine learning: Novel bioinformatics approaches for combating antimicrobial resistance. Current Opinion in Infectious Diseases. 2017;**30**:511-517. DOI: 10.1097/qco.0000000000000406

[18] González-Camacho JM, Ornella L, Pérez-Rodríguez P, Gianola D, Dreisigacker S, Crossa J. Applications of machine learning methods to genomic selection in breeding wheat for rust resistance. The Plant Genome. 2018;**11**. DOI: 10.3835/plantgenome2017.11.0104

[19] Athey S. The Impact of Machine Learning on Economics. In: Agrawal A, Gans J, Goldfarb A, editors. The Economics of Artificial Intelligence: An Agenda. Chicago: University of Chicago Press; 2018

[20] Lesiv M, Schepaschenko D, Moltchanova E, Bun R, Dürauer M, Prishchepov AV, et al. Spatial distribution of arable and abandoned land across former Soviet Union countries. Scientific Data. 2018;**5**:180056. DOI: 10.1038/sdata.2018.56

[21] Maxwell AE, Warner TA, Fang F. Implementation of machine-learning classification in remote sensing: An applied review. International Journal of Remote Sensing. 2018;**39**:2784-2817. DOI: 10.1080/01431161.2018.1433343

[22] Li Y, Zhang H, Xue X, Jiang Y, Shen Q. Deep learning for remote sensing image classification: A survey. Wiley Interdisciplinary Reviews: Data Mining and Knowledge Discovery. 2018;**8**:e1264. DOI: 10.1002/widm.1264

[23] Berry T, Harlim J. Correcting biased observation model error in data assimilation. Monthly Weather Review. 2017;**145**:2833-2853. DOI: 10.1175/mwr-d-16-0428.1

[24] Abbot J, Marohasy J. The application of machine learning for evaluating anthropogenic versus natural climate change. GeoResJ. 2017;**14**:36-46. DOI: 10.1016/j.grj.2017.08.001

[25] Cortes C, Vapnik V. Support-vector networks. Machine Learning. 1995;**20**:273-297. DOI: 10.1023/a:1022627411411

[26] Smola AJ, Schölkopf B. A tutorial on support vector regression. Statistics and Computing. 2004;**14**:199-222. DOI: 10.1023/b:Stco.0000035301.49549.88

[27] Suykens JAK, Van Gestel T, De Brabanter J, De Moor B, Vandewalle J. Least Squares Support Vector Machines. Singapore: World Scientific; 2002

[28] Balabin RM, Lomakina-Rumyantseva E. Support vector machine regression (SVR/LS-SVM) - an alternative to neural networks (ANN) for analytical chemistry– comparison of nonlinear methods on near infrared (NIR) spectroscopy data. The Analyst. 2011;**136**:1703-1712. DOI: 10.1039/c0an00387e

[29] Karasuyama M, Nakano R, editors. Optimizing SVR Hyperparameters via fast cross-validation using AOSVR. In: 2007 International Joint Conference on Neural Networks. 2007

[30] Wang L. Support Vector Machines: Theory and Applications. Berlin Heidelberg: Springer-Verlag; 2005

[31] Sheela KG, Deepa SN. Review on methods to fix number of hidden neurons in neural networks. Mathematical Problems in Engineering. 2013;**2013**:11. DOI: 10.1155/2013/425740

[32] Hinton GE, Osindero S, Teh Y-W. A fast learning algorithm for deep belief nets. Neural Computation. 2006;**18**:

1527-1554. DOI: 10.1162/neco.2006.18.7.1527

[33] Hinton GE, Salakhutdinov RR. Reducing the dimensionality of data with neural networks. Science. 2006; **313**:504-507. DOI: 10.1126/science.1127647

[34] Hinton GE. A practical guide to training restricted Boltzmann machines. In: Montavon G, Orr GB, Müller K-R, editors. Neural Networks: Tricks of the Trade. 2nd ed. Berlin, Heidelberg: Springer Berlin Heidelberg; 2012. pp. 599-619

[35] Breiman L. Bagging Predictors. Machine Learning. Vol. 241996. pp. 123-140. DOI: 10.1023/a:1018054314350

[36] Efron B. Better bootstrap confidence intervals. Journal of the American Statistical Association. 1987; **82**:171-185. DOI: 10.1080/01621459.1987.10478410

[37] Awajan AM, Ismail MT, AL Wadi S. Improving forecasting accuracy for stock market data using EMD-HW bagging. PLoS One. 2018;**13**. DOI: 10.1371/journal.pone.0199582

[38] Breiman L. Random Forests. Machine Learning. Vol. 452001. pp. 5-32. DOI: 10.1023/a:1010933404324

[39] Mayumi Oshiro T, Santoro Perez P, Baranauskas J. How Many Trees in a Random Forest? Lecture notes in computer science 7376 MLDM'12 Proceedings of the 8th international conference on Machine Learning and Data Mining in Pattern Recognition; 2012. pp. 154-168

[40] Liaw A, Wiener M. Classification and regression by randomForest. R News. 2002;**2**:18-22

[41] Jang J. ANFIS: Adaptive-network-based fuzzy inference system. IEEE Transactions on Systems, Man, and Cybernetics. 1993;**23**:665-685. DOI: 10.1109/21.256541

[42] Cheng C-T, Lin J-Y, Sun Y-G, Chau K. Long-Term Prediction of Discharges in Manwan Hydropower Using Adaptive-Network-Based Fuzzy Inference Systems Models. Berlin, Heidelberg: Springer Berlin Heidelberg; 2005. pp. 1152-1161

[43] Ali M, Deo R, Downs N, Maraseni T. An ensemble-ANFIS based uncertainty assessment model for forecasting multi-scalar standardized precipitation index. Atmospheric Research. 2018;**207**:155-180. DOI: 10.1016/j.atmosres.2018.02.024

[44] Şahin M, Rızvan E. Prediction of attendance demand in European football games: Comparison of ANFIS, fuzzy logic, and ANN. Computational Intelligence and Neuroscience. 2018; **2018**:14. DOI: 10.1155/2018/5714872

[45] Schapire RE. The boosting approach to machine learning: An overview. In: Denison DD, Hansen MH, Holmes CC, Mallick B, Yu B, editors. Nonlinear Estimation and Classification. New York, NY: Springer New York; 2003. pp. 149-171

[46] Friedman JH. Greedy function approximation: A gradient boosting machine. The Annals of Statistics. 2001; **29**:1189-1232

[47] Khashei M, Bijari M. A new class of hybrid models for time series forecasting. Expert Systems with Applications. 2012;**39**:4344-4357. DOI: 10.1016/j.eswa.2011.09.157

[48] Zhang GP. Time series forecasting using a hybrid ARIMA and neural network model. Neurocomputing. 2003; **50**:159-175. DOI: 10.1016/S0925-2312(01)00702-0

[49] Sanghani A, Bhatt N, Chauhan NC, editors. A Novel Hybrid Method for

Time Series Forecasting Using Soft Computing Approach. Cham: Springer International Publishing; 2019

[50] Addison PS. The Illustrated Wavelet Transform Handbook. Introductory Theory and Applications in Science, Engineering, Medicine and Finance. Boca Raton, FL: CRC Press; 2017

[51] Najmi A-H, Sadowsky J, Morlet O, Transform W. The continuous wavelet transform and variable resolution time-frequency analysis. Johns Hopkins APL Technical Digest (Applied Physics Laboratory). 1997;**18**:134-139

[52] Sundararajan D. Discrete Wavelet Transform: A Signal Processing Approach. First ed. John Wiley & Sons: Singapore Pte Ltd.; 2015

[53] Mishra A, Singh V. A review of drought concepts. Journal of Hydrology. 2010;**391**:202-216. DOI: 10.1016/j.jhydrol.2010.07.012

[54] Heim Jr RR. A review of twentieth-century drought indices used in the United States. Bulletin of the American Meteorological Society. 2002;**83**: 1149-1166. DOI: 10.1175/1520-0477-83.8.1149

[55] World Meteorological Organization (WMO) and Global Water Partnership (GWP) Integrated Drought Management Programme (IDMP). Handbook of Drought Indicators and Indices. Svoboda M, Fuchs BA, editors 2016

[56] McKee TB, Doesken NJ, Kleist J. The relationship of drought frequency and duration of time scales. In: 8th Conference of Applied Climatology. CA, USA: Anaheim; 1993. pp. 179-184

[57] Hayes M, Svoboda M, Wall N, Widhalm M. The Lincoln declaration on drought indices: Universal meteorological drought index recommended. Bulletin of the American Meteorological Society. 2011;**92**:485-488. DOI: 10.1175/2010BAMS3103.1

[58] R Core Team. R: A Language and Environment for Statistical Computing. Vienna: R Foundation for Statistical Computing; 2013

[59] Beguería S, Vicente-Serrano SM, Reig F, Latorre B. Standardized precipitation evapotranspiration index (SPEI) revisited: Parameter fitting, evapotranspiration models, tools, datasets and drought monitoring. International Journal of Climatology. 2014;**34**: 3001-3023. DOI: 10.1002/joc.3887

[60] Neves J. Package 'spi'. Compute the SPI index using R.; 2011

[61] Byun H-R, Wilhite DA. Objective quantification of drought severity and duration. Journal of Climate. 1999;**12**: 2747-2756. DOI: 10.1175/1520-0442(1999)012<2747:Oqodsa>2.0.Co;2

[62] Bazrafshan J, Khalili A. Spatial analysis of meteorological drought in Iran from 1965 to 2003. Desert. 2013;**18**: 63-71

[63] Tian Y, Nearing GS, Peters-Lidard CD, Harrison KW, Tang L. Performance metrics, error modeling, and uncertainty quantification. Monthly Weather Review. 2016;**144**:607-613. DOI: 10.1175/mwr-d-15-0087.1

[64] Waseem M, Mani N, Andiego G, Usman M. A review of criteria of fit for hydrological models. International Research Journal of Engineering and Technology (IRJET). 2017;**4**:1765-1772

[65] Nash JE, Sutcliffe JV. River flow forecasting through conceptual models part I — A discussion of principles. Journal of Hydrology. 1970;**10**:282-290. DOI: 10.1016/0022-1694(70)90255-6

[66] Willmott CJ. On the validation of models. Physical Geography. 1981;**2**: 184-194. DOI: 10.1080/ 02723646.1981.10642213

[67] Shirmohammadi B, Moradi H, Moosavi V, Semiromi M, Zeinali A. Forecasting of meteorological drought using wavelet-ANFIS hybrid model for different time steps (case study: Southeastern part of East Azerbaijan province, Iran). Natural Hazards. 2013; **69**:389-402. DOI: 10.1007/s11069-013-0716-9

[68] Belayneh A, Adamowski J, Khalil B. Short-term SPI drought forecasting in the Awash River basin in Ethiopia using wavelet transforms and machine learning methods. Sustainable Water Resources Management. 2016;**2**: 87-101. DOI: 10.1007/s40899-015-0040-5

[69] Mishra AK, Desai VR, Singh VP. Drought forecasting using a hybrid stochastic and neural network model. Journal of Hydrologic Engineering. 2007;**12**:626-638. DOI: 10.1061/(ASCE) 1084-0699(2007)12:6(626)

[70] Belayneh A, Adamowski J. Standard precipitation index drought forecasting using neural networks, wavelet neural networks, and support vector regression. Applied Computational Intelligence and Soft Computing. 2012; **2012**:1-13. DOI: 10.1155/2012/794061

[71] Belayneh A, Adamowski J, Khalil B, Quilty J. Coupling machine learning methods with wavelet transforms and the bootstrap and boosting ensemble approaches for drought prediction. Atmospheric Research. 2016;**172**:37-47. DOI: 10.1016/j.atmosres.2015.12.017

[72] Frank E, Wang Y, Inglis S, Holmes G, Witten IH. Using model trees for classification. Machine Learning. 1998; **32**:63-76. DOI: 10.1023/a: 1007421302149

[73] Strohmann T, Grudic G. A formulation for minimax probability machine regression. In: NIPS'02 Proceedings of the 15th International Conference on Neural Information Processing Systems. Cambridge, MA, USA: MIT Press; 2003. pp. 785-792

[74] Khosravi I, Jouybari-Moghaddam Y, Sarajian MR. The comparison of NN, SVR, LSSVR and ANFIS at modeling meteorological and remotely sensed drought indices over the eastern district of Isfahan, Iran. Natural Hazards. 2017; **87**:1507-1522. DOI: 10.1007/s11069-017-2827-1

[75] Chen J, Li M, Wang W. Statistical uncertainty estimation using random forests and its application to drought forecast. Mathematical Problems in Engineering. 2012;**2012**:1-12. DOI: 10.1155/2012/915053

[76] Agana AN, Homaifar A. EMD-based predictive deep belief network for time series prediction: An application to drought forecasting. Hydrology. 2018;**5**. DOI: 10.3390/hydrology5010018

[77] El Shafie AH, El-Shafie A, Almukhtar A, Taha M, El Mazoghi HG, Abou Kheira A. Radial basis function neural networks for reliably forecasting rainfall. Journal of Water and Climate Change. 2012;**3**:125

[78] Sumi SM, Zaman MF, Hirose H. A rainfall forecasting method using machine learning models and its application to the Fukuoka city case. International Journal of Applied Mathematics and Computer Science. 2012;**22**:841-854. DOI: 10.2478/ v10006-012-0062-1

[79] Friedman JH. Multivariate adaptive regression splines. Annals of Statistics. 1991;**19**:1-67

[80] Hastie T, Robert Tibshirani R, Friedman J. The Elements of Statistical

Learning: Data Mining, Inference, and Prediction. New York: Springer; 2009

[81] Javanmard S, Yatagai A, Nodzu M, Bodaghjamali J, Kawamoto H. Comparing high-resolution gridded precipitation data with satellite rainfall estimates of TRMM_3B42 over Iran. Advances in Geosciences. 2010;**25**: 119-125. DOI: 10.5194/adgeo-25-119-2010

[82] Dembélé M, Zwart SJ. Evaluation and comparison of satellite-based rainfall products in Burkina Faso, West Africa. International Journal of Remote Sensing. 2016;**37**: 3995-4014. DOI: 10.1080/01431161.2016.1207258

[83] Harris I, Jones PD, Osborn TJ, Lister DH. Updated high-resolution grids of monthly climatic observations – The CRU TS3.10 dataset. International Journal of Climatology. 2014;**34**: 623-642. DOI: 10.1002/joc.3711

[84] Yatagai A, Kamiguchi K, Arakawa O, Hamada A, Yasutomi N, Kitoh A. APHRODITE: Constructing a long-term daily gridded precipitation dataset for Asia based on a dense network of rain gauges. Bulletin of the American Meteorological Society. 2012;**93**:1401-1415. DOI: 10.1175/bams-d-11-00122.1

[85] Joseph R, Smith TM, Sapiano MRP, Ferraro RR. A new high-resolution satellite-derived precipitation dataset for climate studies. Journal of Hydrometeorology. 2009;**10**:935-952. DOI: 10.1175/2009jhm1096.1

[86] Adler RF, Huffman GJ, Chang A, Ferraro R, Xie P-P, Janowiak J, et al. The version-2 global precipitation climatology project (GPCP) monthly precipitation analysis (1979–present). Journal of Hydrometeorology. 2003;**4**: 1147-1167. DOI: 10.1175/1525-7541(2003)004<1147:Tvgpcp>2.0.Co;2

[87] Kanamitsu M, Ebisuzaki W, Woollen J, Yang S-K, Hnilo JJ, Fiorino M, et al. NCEP–DOE AMIP-II reanalysis (R-2). Bulletin of the American Meteorological Society. 2002;**83**:1631-1644. DOI: 10.1175/bams-83-11-1631

[88] Berrisford P, Dee DP, Poli P, Brugge R, Mark F, Manuel F, et al. The ERA-Interim Archive Version 2.0. Shinfield Park, Reading: ECMWF; 2011

[89] Jian-Jian F, Shuanglin L. Intercomparison of the south Asian high in NCEP1, NCEP2, and ERA-40 reanalyses and in station observations. Atmospheric and Oceanic Science Letters. 2012;**5**:189-194. DOI: 10.1080/16742834.2012.11446989

[90] Maggioni V, Nikolopoulos EI, Anagnostou EN, Borga M. Modeling satellite precipitation errors over mountainous terrain: The influence of gauge density, seasonality, and temporal resolution. IEEE Transactions on Geoscience and Remote Sensing. 2017;**55**:4130-4140. DOI: 10.1109/TGRS.2017.2688998

[91] AghaKouchak A, Mehran A, Norouzi H, Behrangi A. Systematic and random error components in satellite precipitation data sets. Geophysical Research Letters. 2012;**39**. DOI: 10.1029/2012GL051592

[92] Tian Y, Peters-Lidard CD, Eylander JB, Joyce RJ, Huffman GJ, Adler RF, et al. Component analysis of errors in satellite-based precipitation estimates. Journal of Geophysical Research-Atmospheres. 2009;**114**. DOI: 10.1029/2009JD011949

[93] Domenikiotis C, Spiliotopoulos M, Galakou E, Dalezios N, editors. Assessment of the Cold Cloud Duration (Ccd) Methodology for Rainfall Estimation in Central Greece; 2003

[94] Laviola S, Levizzani V. Passive microwave remote sensing of rain from satellite sensors. In: Mukherjee M, editor. Advanced Microwave and

Millimeter Wave Technologies Semiconductor Devices Circuits and Systems. Rijeka: InTech; 2010. pp. 549-572

[95] Kawanishi T, Kuroiwa H, Kojima M, Oikawa K, Kozu T, Kumagai H, et al. TRMM precipitation radar. Advances in Space Research. 2000;**25**:969-972. DOI: 10.1016/S0273-1177(99)00932-1

[96] Bell TL, Kundu PK. Comparing satellite rainfall estimates with rain gauge data: Optimal strategies suggested by a spectral model. Journal of Geophysical Research-Atmospheres. 2003;**108**. DOI: 10.1029/2002JD002641

[97] Sun Q, Miao C, Duan Q, Ashouri H, Sorooshian S, Hsu K-L. A review of global precipitation data sets: Data sources. Estimation, and Intercomparison. 2018;**56**:79-107. DOI: 10.1002/2017RG000574

[98] Beck HE, Vergopolan N, Pan M, Levizzani V, van Dijk AIJM, Weedon GP, et al. Global-scale evaluation of 22 precipitation datasets using gauge observations and hydrological modeling. Hydrology and Earth System Sciences. 2017;**21**:6201-6217. DOI: 10.5194/hess-21-6201-2017

[99] Funk C, Peterson P, Landsfeld MF, Pedreros D, Verdin JP, Rowland J, et al. A Quasi-Global Precipitation Time Series for Drought Monitoring. USGS professional paper 832; 2014

[100] Nguyen P, Shearer EJ, Tran H, Ombadi M, Hayatbini N, Palacios T, et al. The CHRS data portal, an easily accessible public repository for PERSIANN global satellite precipitation data. Scientific Data. 2019;**6**:180296. DOI: 10.1038/sdata.2018.296

[101] Ashouri H, Hsu K-L, Sorooshian S, Braithwaite DK, Knapp KR, Cecil LD, et al. PERSIANN-CDR: Daily precipitation climate data record from multisatellite observations for hydrological and climate studies. Bulletin of the American Meteorological Society. 2015;**96**:69-83. DOI: 10.1175/bams-d-13-00068.1

[102] Joyce RJ, Janowiak JE, Arkin PA, Xie P. CMORPH: A method that produces global precipitation estimates from passive microwave and infrared data at high spatial and temporal resolution. Journal of Hydrometeorology. 2004;**5**:487-503. DOI: 10.1175/1525-7541(2004)005<0487:Camtpg>2.0.Co;2

[103] Xie P, Arkin PA. Analyses of global monthly precipitation using gauge observations, satellite estimates, and numerical model predictions. AMS. 1996; **9**:840-858. DOI: 10.1175/1520-0442(1996)009<0840:Aogmpu>2.0.Co;2

[104] Huffman GJ, Bolvin DT, Nelkin EJ, Wolff DB, Adler RF, Gu G, et al. The TRMM multisatellite precipitation analysis (TMPA): Quasi-global, multiyear, combined-sensor precipitation estimates at fine scales. Journal of Hydrometeorology. 2007;**8**: 38-55. DOI: 10.1175/jhm560.1

[105] Gorelick N, Hancher M, Dixon M, Ilyushchenko S, Thau D, Moore R. Google earth engine: Planetary-scale geospatial analysis for everyone. Remote Sensing of Environment. 2017;**202**: 18-27. DOI: 10.1016/j.rse.2017.06.031

[106] Lokers R, Knapen R, Janssen S, van Randen Y, Jansen J. Analysis of big data technologies for use in agro-environmental science. Environmental Modelling and Software. 2016;**84**: 494-504. DOI: 10.1016/j.envsoft.2016.07.017

[107] Rahlf T. Data Visualisation with R. Springer International Publishing; 2017. 385 p

[108] Wickham H. ggplot2: Elegant Graphics for Data Analysis. New York: Springer-Verlag; 2016. 213 p

Chapter 2

Can Nonlinear Water Pricing Help to Mitigate Drought Effects in Temperate Countries?

Jean-Philippe Terreaux and Mabel Tidball

Abstract

The notion of drought is most often associated with the aridity of landscapes and vegetation. But a green landscape can hide a frequent imbalance between water availability and the quantity necessary to maintain rivers in a suitable state, to satisfy different water needs. This is the case, for example, in the French region called New Aquitaine. Regularly, "drought" crisis committees are set up there to limit water use through administrative constraints, which is technically difficult and costly for many, and with an overall unsatisfactory situation from rural areas to the coast. But in summer, water consumption is mainly due to irrigation. Some water resource managers have consequently set up an original non-linear water pricing system for irrigation to achieve several objectives: above all, to limit water consumption in order to respect a minimum flow rate in rivers, to anticipate water supply-demand imbalances before agricultural plantations are made, to allocate water to the users who value it best, to recover water supply costs, to be transparent and sufficiently simple in its application to be acceptable. In this chapter, we propose to describe one of such original pricing systems, as well as some of its main mathematical properties and its practical interests.

Keywords: drought, irrigation, nonlinear pricing, environment, mathematical economics

1. The problem of chronic droughts in France

Economic and trade developments and the growth of the world's population are leading to an increasing exploitation of natural resources. In addition, this global evolution is accompanied by changes in the economic, social, and environmental context (climate change and spread of species becoming invasive), making more uncertain the context in which different professions work. For example, the farmers now face a multiplicity of new climatic, market, and biotic risks. In this context, the implementation of measures to secure various resources, including water, helps to mitigate the negative effects of these developments. More generally, various tools to combat hazards have been put in place at the individual level: insurance or other risk pooling systems, diversification, etc.

But with regard to water resources, even in some areas where they were once considered abundant, the limits of their exploitation may have been reached. And because property rights over these resources are often poorly established, the "tragedy of the commons" (see [1]) is repeated. For example, in the agricultural sector, everyone is encouraged to make the most of available water in their own interest, while well-designed coordination of the different uses would improve everyone's well-being or economic performance (see, for example, [2]).

The problems thus created have repercussions on the agricultural activity itself, which can no longer rely on a sufficient supply of water, and on the community in general: on the environmental level (e.g., river dewatering), on the economic level (impact on downstream sectors, such as shellfish farming or tourism, as the European Coastal project shows, see [3]), or on the social level with farmers quite often being accused of many wrongdoings. This is also one of the reasons for the implementation at European Union level of the Water Framework Directive (see e.g., [4]).

Thus, a green landscape such as those frequent in the French region of New Aquitaine can hide recurrent problems of drought, with very frequent crises in areas such as the Charente "département." As a result, the public authorities regularly intervene, imposing different constraints to farmers or to their Water User Associations (WUA) in charge of the resource management (e.g., restrictions on irrigation, while crops are in place) to allocate the water shortage. The situation is then penalizing for everyone, and especially for farmers, who too often cannot adequately grow the crops in which they have invested. And it ultimately leads to a perfectible situation on each of the three pillars of a sustainable development: economic, social, and environmental.

This situation is well documented and explained in game theory, particularly in the context of the "prisoner's dilemma" (see [5]). But it gives few solutions, except to seek coordination between stakeholders. The "tragedy of the commons" can generally be resolved by privatizing the resource, but in France and in many countries, this solution cannot be applied to the water resource for legal reasons (see, for example, [6] for their developments on this subject). This is what opponents of water storage in reservoirs remind us, arguing that the resource must benefit all uses and that it cannot be used by a single sector of the economy.

On the other hand, at some short distance from this Charente département, and still in the New Aquitaine Region, an original attempt at water pricing has been made in order to try to better coordinate farmers' actions, by anticipating as well as possible the annual imbalances between water supply and demand. The objective is thus to make the best use of public information (the water level in rivers, groundwater or reservoirs, climatic conditions, market conditions, etc.) and also private information (importance for farmers of securing their water supply, linked, for example, to their product sales contracts, their debt level, or more simply their risk aversion), but without being inquisitorial (by making them reveal only what is important for the water resource management). Of course, this does not guarantee that there will be no crisis, but it does make it possible to anticipate them as well as possible and to resolve a majority of possible conflicts well before the plots are planted, that is, before the distribution of the shortage has any serious consequences. This too gives the farmers the possibility to change their culture choices in order to adapt them more precisely to the available water.

The idea, developed by the irrigators of the Compagnie d'Aménagement des Eaux des Deux-Sèvres (CAEDS), is to base the price of irrigation water paid each year on two variables: the quantity of water reserved by each farmer (before planting) and the quantity actually consumed. We show here that the pricing formula used encourages the farmer to subscribe a quantity directly related to what he expects to consume (which makes it possible to deduce very quickly what each farmer plans to consume and the total consumption, which makes it easier, as we saw, to resolve possible conflicts, as well as consequently to respect more easily the minimum flow rates in the nearby rivers). Finally, we show that it is possible to modify the volumes consumed globally by varying the parameters.

In a way, by creating value not only from the water resources but also simultaneously from public and private information, this makes it easier to resolve conflicts of use, by reducing the frequency of crises. This approach combines mechanism design, game theory, and nonlinear pricing results.

In the rest of this chapter, we give a very short overview of some of the literature on irrigation water resource pricing (see [7] for some aspects), as well as on nonlinear pricing (see [8]). Then we develop a mathematical model to show different properties of this pricing formula and in particular why volume subscription by farmers makes it possible to anticipate their consumption.

2. Literature review: agricultural water economics and nonlinear pricing

In France, after having built many individual or collective dams in order to increase water-storage capacity ("supply management"), efforts are currently focused on "demand management," that is, the use of less irrigation water for the same production and the search for more efficient alternatives for sharing water among the different uses while trying to find more efficient water pricing schemes (see [9]). The problem is similar in other parts of the world (see, e.g., [10] for China).

The aims of water management are multiple and may sometimes be understood as contradictory ([11]): the first one is to allocate water to users who valorize it at the best (efficiency). The second is to guarantee an access to this essential good to everybody and to be acceptable in order to be applied (equity). Moreover, as mentioned in [12], it may be a tool to redistribute public investment benefits. The third is to recover costs induced by water extraction/distribution/use. The fourth is to be transparent and simple enough to be understandable, and it is clear that a two variable tariff, as the one presented here, is quite acceptable as shown by the fact that it was implemented in the CAEDS area. Another nonlinear pricing scheme is also established in the Compagnie d'Aménagement des Coteaux de Gascogne area in the southwest of France and is compared in [13] and [14] to the one presented here as regards the agricultural production, the farmers revenue, and the water quantity used for irrigation in accordance with the climatic conditions.

Generally speaking, water pricing practices can be classified in two families: volumetric and non-volumetric methods. Volumetric methods rely on the volume and require metered water facility (see [15] or [16], for examples of the implementation of a volumetric pricing system). Non-volumetric methods are based on output/input other than water, for example, in the agricultural sector as per area pricing (see [17]). The last methods are widespread because of their simplicity, but they do not encourage saving water.

When using volumetric methods, water can be priced in three main ways. The price can be either constant whatever the level of consumed water or defined "per block": the cost per additional consumed unit varies when the consumption reaches some given thresholds. The marginal pricing can either increase with the level of consumption (increasing block tariff) or decrease (declining block rate). The application of such a pricing is studied for domestic water use, for example, in [18].

The increasing block tariffs (IBT) can be used to impose conservation incentives on some target group of large users. Customers facing the higher prices at the margin will, in theory, use less water than they would under the uniform pricing; customers facing lower prices at the margin will use more. The expectation is that demand in the high blocks will be more elastic than demand in the low blocks, resulting in a net decrease in water use when compared to a uniform pricing. Although there is widespread consensus that IBT have many advantages, this type of tariff still deserves more careful examination since an incorrect structure of the IBTs leads to several shortcomings as argued in [19]. Some of them are difficulties to set the initial block; mismatch between prices and marginal costs; conflict between revenue sufficiency and economic efficiency; absence of simplicity, transparency, and implementation; incapacity of solving shared connections; etc.

The decreasing block tariff (DBT) is, unlike the preceding one, in accordance with the proposition that high-value goods "should" be bought at higher price than low-value goods. Water will be first purchased for uses with high values and then only for uses which will lead to less welfare increases. Concerning equity, this type of tariff is "not advisable". "The consumers who acquire smaller amounts of the good and/or service because of their low incomes would be bearing a higher price than those who can afford to consume greater amounts" (see [20]). But it can be justified in the following circumstances:

- When users have very different levels of consumption. A consumer hundred times bigger than the average consumer does not create costs hundred times higher, because there is only one pipe line, one billing process, etc. And, since cost per volume is lower with large consumers, it is justifiable to propose DBT in the case of heterogeneous users.

- In order to incite users to stay in the WUA: as we have explained above, IBT (e.g., see [21]) might encourage users who have access to alternative water sources to quit (partly at least) the network, stopping to contribute to the recovery of the costs. This can lead to cost recovery problems for the water supplier and besides might lead to negative environmental consequences. DBT does not have this negative incentive.

A two-part tariff combines a fixed and a volumetric rate (or a mix of fixed and variable elements). "Under this system, consumers must pay an entry charge that entitles them to consume the good. Subsequently they will pay an additional smaller amount for each extra unit consumed." "Two-part tariff is easy to explain and easy to understand" is mentioned in [20]. But in practice, it fails to reach the efficiency objective and suffer from the fact that it does not allow to reveal information on water demand, which may be at the origin of sudden discrepancy between water supply and demand.

In the following sections, we study the properties of a different pricing structure, in which farmers make a water reservation (e.g., during wet period or before planting) and then pay a water bill which is an increasing function of water reservation and of real water consumption (e.g., during dry period or during peak vegetation). This allows the WUA manager to forecast disequilibrium between water demand and supply. The water pricing is parameterized, in order to adapt the price to the actual WUA situation and to the available water supply.

3. The model

3.1 Notations

We consider a WUA composed of n farmers and which provides them irrigation water at a cost. Each farmer i has a production function we note $h_i(C_i)$ which is a function of the volume C_i of the water he consumes. This production function is private information, known only by the farmer himself.

Each year, each farmer firstly reserves a water volume S_i, for example, before choosing his planting and then consumes another volume C_i for the field irrigation, C_i being either inferior or superior to S_i. The pricing formula is designed in order to take into account these two variables and to display some properties.

The notations we use are the following:

- B is the total water user association expenses.
- D is proportional to B: $D = \lambda B$, with a constant $\lambda > 0$.
- S_i is the volume reserved by agent i during the considered year.
- C_i is the volume consumed by agent i during the same year.
- $h_i(C_i)$ is agent i's production function, which depends on the consumed water C_i.
- $F = F(S_i, C_i)$ is the sum agent i must pay (his water bill).

For each agent i, the pricing formula is

$$F(S_i, C_i) = D\left(aS_i + (1-a)\frac{C_i max(C_i, bS_i)}{S_i}\right) \tag{1}$$

with $a \in [0, 1]$ and $b \in [0, 1]$.

The pricing scheme is a common knowledge for all farmers and is the same for all of them. Parameter a represents a kind of sharing of the price between, on the first hand, the reservation part and, on the other hand, the consumption part. As $D = \lambda B$, the role of parameter λ is to ensure a balanced budget, under the financial conditions of the WUA, for example, by adjusting the value of this parameter by trial and error year after year, if a temporary budget imbalance is permissible. We will return to this point in Section 3.3, examining in particular the case where a minimum of revenue each year is required. The role of parameter b is to incite to reserve at least the forecasted consumption divided by b. For a S_i given, when $C_i > bS_i$, the C_i^2 which appears in the pricing formula incites to diminish water consumption.

A deterministic approach, without acquisition of information between the reservation date and the consumption date, is sufficient in order to study some of the properties of this pricing. Of course other properties directly linked to stochastic variables (as the climate) cannot be studied here and are the object of further researches.

But we must note that since C_i depends on S_i, we must take this relationship into account in optimizing the volumes reserved and consumed by the farmer i.

3.2 The maximization problem of farmer i

When choosing the values of his control variables S_i and C_i, the farmer must decide of the optimal value of C_i knowing the optimal value of S_i previously announced. Therefore each farmer must solve

$$\max_{S_i}\left[\max_{C_i}\left(h_i(C_i) - F(S_i, C_i)\right)\right] \tag{2}$$

where the production function of farmer i, h_i, is an increasing and concave function.

3.2.1 The maximization problem in C_i

We note $G(S_i, C_i) = h_i(C_i) - F(S_i, C_i)$, and we calculate the solution of the maximization of $G(S_i, C_i)$ as a function of C_i. We have

$$\frac{\partial G_i(S_i, C_i)}{\partial C_i} = \begin{cases} h_i'(C_i) - 2(1-a)D\dfrac{C_i}{S_i} & \text{if } bS_i < C_i \\ h_i'(C_i) - (1-a)bD & \text{if } bS_i > C_i \end{cases} \tag{3}$$

- For $bS_i < C_i$, we note $C_i^-(S_i)$ the solution in C_i of

$$h_i'(C_i) = 2(1-a)D\frac{C_i}{S_i} \tag{4}$$

With a simple derivation of this last equation, it is easy to see that

$$C_i^{-\prime}(S_i) = \frac{2(1-a)DC_i}{2(1-a)DS_i - h''(C_i)S_i^2} > 0 \tag{5}$$

Therefore $C_i^{-'}(S_i)$ is an increasing function of S_i.
We remark by the way that when $bS_i < C_i$, we have

$$2(1-a)D\frac{C_i}{S_i} > 2(1-a)bD > (1-a)bD \tag{6}$$

- For $bS_i > C_i$, we call C_i^+ the solution in C_i of

$$h_i'(C_i) = (1-a)bD \tag{7}$$

We note that this solution does not depend on S_i.
The preceding remark, the concavity of h_i, and the definition in (4) and (7) of C_i^- and of C_i^+ imply that

$$C_i^-(S_i) < C_i^+ \tag{8}$$

We can easily deduce that the optimal solution C_i^{sol} of (2) for S_i given depends on the relative positions of $C_i^-(S_i) < C_i^+$ and of bS_i:

$$C_i^{sol} = \begin{cases} C_i^-(S_i) & \text{if } S_i < C_i^-(S_i)/b \\ bS_i & \text{if } C_i^-(S_i)/b \le S_i \le C_i^+/b \\ C_i^+ & \text{if } S_i > C_i^+/b \end{cases} \tag{9}$$

Note that this formula gives a relation $C_i^{sol}(S_i)$ that depends on the parameters of the regulator and of the profit function h_i of farmer i. When starting from small S_i,

$C_i^{sol}(S_i)$ is first an increasing function of S_i, then it is a linear function in S_i, and finally it is a constant.

3.2.2 The maximization problem in $\mathrm{S_i}$

When solving the maximization of our problem in S_i, knowing the optimal value C_i, which is generally, as we saw, a function of S_i, we must consider the relation between these two variables.

- If $S_i < C_i^-(S_i)/b$, we must solve

$$\max_{S_i}\left(h_i\left(C_i^-(S_i)\right) - F\left(S_i, C_i^-(S_i)\right)\right) \tag{10}$$

and the first-order condition gives

$$h_i'(C_i^-(S_i))C_i'^-(S_i) = D\left[a + (1-a)\frac{2C_i^-(S_i)C_i'^-(S_i)S_i - \left(C_i^-(S_i)\right)^2}{S_i^2}\right] \tag{11}$$

- If $C_i^-(S_i)/b \le S_i \le C_i^+/b$ we must solve

$$\max_{S_i}\left(h_i(bS_i) - D\left(aS_i + (1-a)b^2S_i\right)\right) \tag{12}$$

for which the first-order condition is

$$bh_i'(bS_i) = D\left[a + (1-a)b^2\right] \tag{13}$$

- and if $S_i > C_i^+/b$ the maximization problem is:

$$\max_{S_i}\left(h_i(C_i^+) - F(S_i, C_i^+)\right) \tag{14}$$

that does not have a solution, which means that the farmer will at least consume bS_i (as consuming less would decrease his production without decreasing his water bill; in other words we have $bS_i \le C_i$).

To obtain the optimal solution of our problem in S_i, we must analyze the admissibility of solutions of (11) and (13).

Theorem

The optimal strategy of farmer i is

1/$S_i = C_i\sqrt{(1-a)/a}$ if $0 < \frac{b^2}{1+b^2} < a < 1$, where C_i is the solution of

$$h_i'(C_i) = 2D\sqrt{a(1-a)} \tag{15}$$

2/$S_i = C_i/b$ if $0 < a < \frac{b^2}{1+b^2} < 1$, where C_i is the solution of

$$h_i'(C_i) = \frac{D\left(a + (1-a)b^2\right)}{b} \tag{16}$$

Proof of the theorem

We start by introducing a Lemma.

Lemma

$$\max_{S_i}\left[\max_{C_i}\left(h_i(C_i) - F(S_i, C_i)\right)\right] = \max_{S_i, C_i}\left[h_i(C_i) - F(S_i, C_i)\right] \tag{17}$$

Proof of the Lemma

In Section 3.2.1 of this chapter, we have shown that the solution of

$$\max_{S_i}\left[\max_{C_i}\left(h_i(C_i) - F(S_i, C_i)\right)\right] \tag{18}$$

is given either if $bS_i = C_i$ (border solution) by the maximization in S_i of $G(S_i, C_i)$ or if $bS_i < C_i$ (interior solution) by $h_i'(C_i) = 2(1-a)D\frac{C_i}{S_i}$ (Eq. (4)).

Now if we solve the problem

$$\max_{S_i, C_i}\left[h_i(C_i) - F(S_i, C_i)\right] \tag{19}$$

its interior solution (with $bS_i < C_i$), knowing that (the first equivalence being due to the fact that $G(S_i, C_i)$ depends on S_i only through $F(S_i, C_i)$ and not through h_i)

$$\frac{\partial G(S_i, C_i)}{\partial S_i} = 0 \Leftrightarrow \frac{\partial F(S_i, C_i)}{\partial S_i} = 0 \Leftrightarrow S_i = \sqrt{\frac{1-a}{a}}\, C_i \text{ and } b\sqrt{\frac{1-a}{a}} < 1 \tag{20}$$

is given by

$$\frac{\partial G(S_i, C_i)}{\partial S_i} = 0 \Leftrightarrow h'(C_i) = 2(1-a)D\frac{C_i}{S_i} \tag{21}$$

which coincides with (4). Replacing this equation in (11), we obtain

$$2(1-a)D\frac{C_i}{S_i}C_i' = D\left(a + \frac{(1-a)2C_iC_i'}{S_i} - (1-a)\frac{C_i^2}{S_i^2}\right) \tag{22}$$

Simplifying we find Eq. (18) so that the interior solution coincides with the solution of (2). The border solutions are also the same (i.e., $bS_i = C_i$) for (2) and for (19). Finally, both solutions are the same.

We now return to the demonstration of the Theorem itself. Thanks to the Lemma, we can now compute the solution of (2), by computing the solution of

$$\max_{S_i, C_i}\left[h_i(C_i) - F(S_i, C_i)\right] \tag{23}$$

- If $bS_i < C_i$,

then (23) can be written as

$$\max_{S_i, C_i}\left(h_i(C_i) - D\left(aS_i + (1-a)\frac{C_i^2}{S_i}\right)\right) \tag{24}$$

And the first order conditions give

$$\text{a/}\ \frac{\partial\left[h_i(C_i) - D\left(aS_i + (1-a)\frac{C_i^2}{S_i}\right)\right]}{\partial S_i} = a - \frac{(1-a)C_i^2}{S_i^2} = 0 \tag{25}$$

which is equivalent to

$$S_i = \sqrt{\frac{1-a}{a}}C_i \tag{26}$$

We remark that this implies, as $bS_i < C_i$, that

$$b\sqrt{\frac{1-a}{a}} < 1 \tag{27}$$

and

$$\text{b/}\ \frac{\partial\left[h_i(C_i) - D\left(aS_i + (1-a)\frac{C_i^2}{S_i}\right)\right]}{\partial C_i} = h'(C_i) - D\frac{(1-a)2C_i}{S_i} = 0 \tag{28}$$

which is equivalent to

$$h'(C_i) = 2D\sqrt{a(1-a)} \tag{29}$$

- If $bS_i \geq C_i$,

then (23) can be written as

$$\max_{S_i,\, C_i} \left(h_i(C_i) - D(aS_i + (1-a)bC_i)\right) + \mu(bS_i - C_i) \tag{30}$$

with μ the dual variable associated to the constraint $bS_i \geq C_i$.
The first-order conditions give here:

$$\text{a/}\ \frac{\partial[h_i(C_i) - D(aS_i + (1-a)bC_i) + \mu(bS_i - C_i)]}{\partial S_i} = -Da + \mu b = 0 \tag{31}$$

and

$$\text{b/}\ \frac{\partial[h_i(C_i) - D(aS_i + (1-a)bC_i) + \mu(bS_i - C_i)]}{\partial C_i} = h'(C_i) - D(1-a)b - \mu = 0 \tag{32}$$

The solution of these equations is.

$$C_i = bS_i,\ \mu = \frac{Da}{b} > 0 \text{ and } h'(C_i) = D\left[(1-a)b + \frac{a}{b}\right] = D\left[\frac{(1-a)b^2 + a}{b}\right] \tag{33}$$

Remark 1
The optimal solution is a continuous function of parameters a and b. Moreover, the regulator can choose parameters a and b in order to enforce an interior solution or a border solution.

Remark 2

Note that in 2/ of the theorem

$$\lim_{b \to 0} \left(\frac{D(a + (1-a)b^2)}{b} \right) = \infty \tag{34}$$

If we choose b small enough and therefore a small enough to remain in case 2/, (20) incites the farmers to use less water, that is, $C_i \to 0$ when $b \to 0$. But in general we cannot draw any conclusion on the value of S_i.

In conclusion the WUA manager may use these two parameters a and b in order to decrease the water consumption, but he cannot make water decrease at discretion since as in our example he might decrease also the reserved volume and at the end the budget equilibrium would not be satisfied.

Note also that in 1/ of the Theorem, we cannot make the consumption C_i decrease at will, since the maximum value of $h_i'(C_i)$ is equal to D according to Eq. (15).

Figure 1 shows how S_i must be tightly correlated to C_i by the farmer in order to obtain a good remuneration $G_i(S_i, C_i) = h_i(C_i) - F(S_i, C_i)$for his activities. (Numerically, it is computed with the following functions and values: $h_i(C_i) = 2.C_i^{0.5}$; a = 1/3; b = 0.7; D = 2; negative values have been replaced by 0). A slight deviation from the optimum value of S_i at the reservation time, and of the optimal consumption C_i, once S_i is chosen, will diminish considerably the value of the gain G. This means that from the value of S_i, the WUA manager is able to predict accurately the level of the water demand.

3.3 The budget equilibrium constraint

In this section, we study the conditions in which the budget equilibrium may be obtained, or in other terms, in which

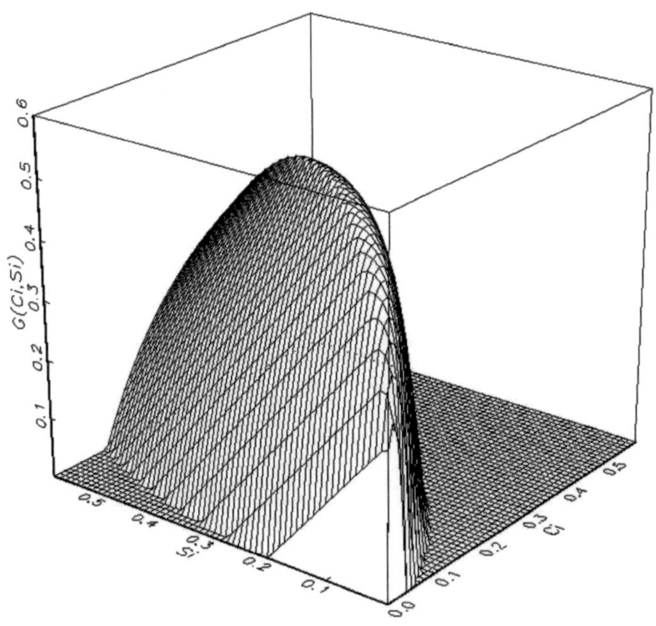

Figure 1.
Representation of $G(C_i, S_i)$ as a function of C_i and of S_i. Numerical values: see text.

$$\sum_i F(S_i, C_i) = B \tag{35}$$

The WUA manager may choose parameters a and b in such a way that

1/ either $0 < \frac{b^2}{1+b^2} < a < 1$ and then $S_i = C_i\sqrt{(1-a)/a}$ according to the Theorem; we know then that

$$C_i = \left[h_i'\right]^{-1}\left(2D\sqrt{a(1-a)}\right) =: g_i\left(2D\sqrt{a(1-a)}\right) \tag{36}$$

and the budget equilibrium constraint (35) must be written as

$$2\lambda B\sqrt{(1-a)/a}\sum_i C_i = 2\lambda B\sqrt{(1-a)/a}\sum_i g_i\left(2\lambda B\sqrt{a(1-a)}\right) = B \tag{37}$$

which gives

$$2\lambda\sqrt{(1-a)/a}\sum_i g_i\left(2\lambda B\sqrt{a(1-a)}\right) = 1 \tag{38}$$

If we assume in order to facilitate the presentation of the demonstration that $h_i(C_i) = \frac{C_i^{\alpha_i}}{\alpha_i}$, with $0 < \alpha_i < 1$, which reminds us of a Cobb–Douglas production function, this last equation becomes

$$f(\lambda) =: 2\sqrt{(1-a)/a}\sum_i \lambda^{\frac{\alpha_i}{\alpha_i - 1}} M_i = 1 \tag{39}$$

$$\text{where } M_i =: \left(2B\sqrt{a(1-a)}\right)^{\frac{\alpha_i}{\alpha_i-1}} \tag{40}$$

Noting that since $\frac{\alpha_i}{\alpha_i - 1} < 0$, we have $\lim_{\lambda\to 0} f(\lambda) = +\infty$ and also $\lim_{\lambda\to\infty} f(\lambda) = 0$. As $f'(\lambda) < 0$, we deduce that there exists a unique λ which verifies (39).

2/ or $0 < a < \frac{b^2}{1+b^2} < 1$ and then $bS_i = C_i$, according to the Theorem.
Previously we showed that

$$C_i = \left[h_i'\right]^{-1}\left(\frac{D\left(a + (1-a)b^2\right)}{b}\right) =: k_i\left(\frac{D\left(a + (1-a)b^2\right)}{b}\right) \tag{41}$$

Assuming here too that $h_i(C_i) = \frac{C_i^{\alpha_i}}{\alpha_i}$, the budget equilibrium constraint (39) can be written as

$$g(\lambda) =: A\sum_i \lambda^{\frac{\alpha_i}{\alpha_i - 1}} N_i = 1 \tag{42}$$

with $A =: \frac{D\left(a+(1-a)b^2\right)}{b}$ and $N_i =: (BA)^{\frac{1}{\alpha_i - 1}}$, and we obtain the same conclusion as in 1/.

So, once parameters a and b are chosen for considerations of water savings, the WUA manager can force the system to be in budgetary equilibrium with the choice of the parameter λ value. Of course, not knowing the true value of α_i parameters, or more generally ignoring the precise form of the $h_i(C_i)$ functions, he will not be able to compute directly the optimal value of λ, but the existence result on a unique λ value and the monotonicity of $f(\lambda)$ and of $g(\lambda)$ allows him to find the correct value by trials and errors.

4. Conclusion

We have shown here how it is possible with a pricing system based on two variables, reservation and consumption, for the WUA manager to get enough information in order to anticipate any disequilibrium between water demand and supply, when it is always possible to change the choice of the cultures. Moreover changing the parameters allows the WUA manager to modify the volume consumed by the farmers, which is especially useful when searching a decrease of the water consumption. Translated in a two-entry table, this method is simple enough to be understood by each farmer and quite acceptable since associating the pursuit of fairness, efficiency, and adaptability.

At last, with a judicious choice of the value for the parameters, it is possible to incite the farmers to be more or less acute in the choice of their reservation and consumption values. This pricing system should therefore allow a more efficient use of the water resource by the farmers, by the way decreasing the constraints on other economic sectors and on the environment.

The need for a better management of irrigation water is now recognized, and the potentialities of original pricing systems (see, for example, [22]) are confirmed by many work in economics, carried out in different contexts such as those presented by [23] in semi-arid climates, or [24], and [25] in different European countries.

Further researches are nevertheless needed to study how such a system keeps or increases its advantages when we take into account the fact that in many countries the water supply may be stochastic (see, for example, [26]). It would also be important to study the strategic interactions between farmers, as well as the different inter-annual dynamics that can be put in place, in order to facilitate the development of agricultural activities, while still under budgetary constraint (see [27]). In addition, the acquisition of information between the reservation (during the wet vegetation season) and the peak consumption (during the dry vegetation season) can be sequential. Taking this into account can lead to an even better valorization of the water resources, through the implementation of an adapted pricing policy. This leads to other refinements which are the aim of other present researches.

Acknowledgements

We thank especially the financial help the European Union H2020 COASTAL project (Grant Agreement No. 773782) and of French Agence Nationale pour la Recherche program" Agriculture et développement durable" allocated to the" Appeau" contract. We thank too Jean-Marc Berland and Jean-Antoine Faby from the International Office for Water for interesting discussions on a preliminary version of this paper. We also thank Laboratoire Montpelliérain d'Economie Théorique et Appliquée (Lameta), now Center for Environmental Economics - Montpellier (CEEM), for allowing us to have previously made available the working paper on which this chapter is partially based (see [28]).

Conflict of interest

The authors declare that there is no conflict of interest.

Author details

Jean-Philippe Terreaux[1]* and Mabel Tidball[2]

1 UR ETBX, Irstea, Cestas, France

2 CEE-M, University of Montpellier, CNRS, Inra, SupAgro, Montpellier, France

*Address all correspondence to: jean-philippe.terreaux@irstea.fr

IntechOpen

© 2019 The Author(s). Licensee IntechOpen. This chapter is distributed under the terms of the Creative Commons Attribution License (http://creativecommons.org/licenses/by/3.0), which permits unrestricted use, distribution, and reproduction in any medium, provided the original work is properly cited. CC BY

References

[1] Hardin G. The tragedy of the commons. Science. 1968;**162**(3859): 1243-1248

[2] Howe CW. The development of an efficient water market in northern Colorado, USA. In: Lago M et al., editors. Use of Economic Instruments in Water Policy. Switzerland: Springer; 2015. pp. 301-315

[3] Tiller R. Coastal–Collaborative Land-Sea Integration Platform, Deliverable D03, WP1 – T1.1 Sector Analysis of Coastal and Rural Development. Sintef, Trondheim; 2019. 124p

[4] Artell J, Huhtala A. What are the benefits of the water framework directive? Lessons learned for policy design from preference revelation. Environmental and Resource Economics. 2017;**68**(4):847-873

[5] Tucker A. A two-person dilemma. Stanford University, mimeo. May 1950. Reprinted in Straffin P. the Prisoner's dilemma. The UMAP Journal. 1980;**1**: 101-103. (unpublished)

[6] Chebly JE. The value of water: Economics of water for a sustainable use. The Economic and Social Review. 2014;**45**(2):207-222

[7] Lago M, Mysiak J, Gomez CM, Delacamara G, Maziotis A, editors. Use of Economic Instruments in Water Policy, Insights from International Experience. Germany: Springer-Verlag; 2015. 423p

[8] Wilson RB. Nonlinear Pricing. Oxford: Oxford University Press; 1993. 429p

[9] Garcia S, Reynaud A. Estimating the benefits of efficient water pricing in France. Resource and Energy Economics. 2004;**26**:1-25

[10] Feike T, Henseler M. Multiple policy instruments for sustainable water management in crop production—A modeling study for the Chinese Aksu-Tarim region. Ecological Economics. 2017;**135**:42-54

[11] Johansson RC, Tsur Y, Toe TL, Doukkali R, Dinar A. Pricing irrigation water: A review of theory and practice. Water Policy. 2002;**4**(2):173-199

[12] Perry C. Water at any price? Issues and options in charging for irrigation water. Irrigation and Drainage. 2001;**50**: 1-7

[13] Sidibé Y, Terreaux JP, Tidball M, Reynaud A. Coping with drought with innovating pricing systems: The case of two irrigation water management companies in France. Agricultural Economics. 2012;**43**(Suppl. 1):141-155

[14] Terreaux JP, Sidibé Y, Tidball M. Préservation des étiages par des méthodes originales de tarification de l'eau d'irrigation. La Houille Blanche. 2012;**6**:41-46

[15] Kajisa K, Moya P, Valencia MS, Valentin MRS, Garcia L. Lessons from volumetric water pricing trials at the three surface irrigation systems in northern Luzon, Philippines. DLSU Business and Economics Review. 2018; **28**:51-69. (Special issue)

[16] Kajisa K, Dong B. The effects of volumetric pricing policy on farmers' water management institutions and their water use: The case of water user organization in an irrigation system in Hubei, China. World Bank Economic Review. 2017;**31**(1):220-240

[17] Johansson RC. Pricing Irrigation Water: A Literature Survey, Policy Research Working Paper. Washington DC: The World Bank; 2000. 87p

[18] Sahin O, Bertone E, Beal CD. A systems approach for assessing water conservation potential through demand-based water tariffs. Journal of Cleaner Production. 2017;**148**:773-784

[19] Boland JJ, Whittington D. The political economy of water tariff design in developing countries: Increasing block tariffs versus uniform price with rebate. In: Dinar A, editor. The Political Economy of Water Pricing Reforms. Oxford: Oxford University Press; 2000. pp. 215-235

[20] Gracia FA, Garcia Valinas MA, Martinez-Espineira R. The Literature on the Estimation of Residential Water Demand. St. Francis University, Department of Economics Working Paper Series; 2001. 5p

[21] Miyawaki K, Omori Y, Hibiki A. Exact estimation of demand functions under block-rate pricing. Econometric Reviews. 2016;**35**(3):311-343

[22] Terreaux JP, Tidball M. Water sharing among competing farmers in temperate climate: A study of different pricing mechanisms. In: World Water Congress; Montpellier. 1st-4th September 2008. 14p

[23] Tidball M, Terreaux JP, Faby JA, Berland JM, Feinerman E, Axelrad G. Efficient allocation of water resources among competing users: Economic, Environmental and organizational consideration. In: Final Report. High Council for Science and Technology Research, Status Seminar on Agronomy and Sustainable Development. Jerusalem; 2008. 62p

[24] Sidibé Y, Terreaux JP, Tidball M, Fernandez D, Mayenco MU, Berland JM, et al. Noviwam: Five European countrics together for improving water quality and water use efficienc. In: XIV World Water Congress. Porto de Galinhas. Brasil; 25-29 September 2011. 7p

[25] Sidibé Y, Terreaux JP, Tidball M, Berland JM, Jacquin N. Le projet européen Noviwam et la gestion intégrée de l'eau. Revue des Sciences de l'Eau. 2016;**29**(2):91-94

[26] Sidibé Y, Terreaux JP, Tidball M. Tarification différentielle de l'eau d'irrigation en cas de ressources aléatoires. In: 4ièmes journées de recherches en sciences sociales. Rennes: Agro-Campus Ouest; 9-10 décembre 2010. 23p

[27] Sidibe Y, Terreaux JP, Tidball M. Innovative management systems to cope with drought: The case of south-western France. In: Symposium on Data Driven Approaches to Drought; Purdue University, West Lafayette, Indiana; 21-22 June; 2011. 8p

[28] Tidball M, Terreaux JP. Information revelation through irrigation water pricing using volume reservations. Lameta Working Paper; DR 2008–14; 2008. 17p

Chapter 3

Drought Impacts on Bioenergy Supply System Risk and Biomass Composition

Amber Hoover, Rachel Emerson, Jason Hansen, Damon Hartley and Allison Ray

Abstract

Bioenergy is an important renewable energy option worldwide, but the industry is susceptible to a myriad of risks including biomass supply, of which drought plays a role. Crops yields decrease during drought, increasing year-to-year risk for the agricultural industry. For the renewable energy industry, in particular, the effect of drought on crops is substantial and complex. This chapter discusses the current state of knowledge regarding how drought affects biomass destined for renewable energy as it relates to dry biomass yields and chemistry, the latter of which heavily impacts cost of production and final product yields. Advanced supply systems are one option for reducing biomass supply risk. These systems lead to higher, less variable crop yields during uncontrollable events like drought; however, the quality of material supplied in a drought year may still vary as drought impacts plant chemistry. This chapter provides analysis for chemical composition of four bioenergy crops observing that both carbohydrates and lignin decrease during a drought year compared to a year with minimal to no drought. These chemical changes can impact biochemical conversion through inhibitor formation and altering degradability during pretreatment.

Keywords: bioenergy, drought, chemical composition, inhibitors, yield, supply system, biorefinery

1. Introduction

Bioenergy is one of a portfolio of renewable energy options used worldwide to support efforts to decrease use of fossil fuels and support energy security policies. By 2050, the total world bioenergy potential is predicted to meet 25–33% of the world's energy demand [1]. One study estimates that in the U.S., by 2040, more than 1 billion tons of biomass could be available for use in the bioenergy industry; however, the water consumption necessary to support these crops is a clear concern, and recent analyses investigate scenarios with purpose grown energy crops that are assumed to be rain fed rather than irrigated [2]. Energy crops are an important strategy for the emerging bioenergy industry, but erratic environmental factors remain a risk with drought being a major factor affecting crop production, particularly for crops grown without irrigation. Widespread droughts covering 30% of the U.S. have occurred every decade since 1900, and drought frequency has increased

IntechOpen

in recent decades [3, 4]. To make matters worse, extreme weather events, like drought, are predicted to become more prevalent under future climate scenarios with corresponding decreases in gross primary productivity [5–8]. The economic impacts of drought are exemplified by the $30 billion in losses from a recent U.S. nationwide drought in 2012 that primarily impacted the agricultural industry as a result of outcomes such as a 27% reduction in U.S. corn grain yields [9]. These yield losses pose considerable risk for biomass producers and biorefineries that already have substantial startup challenges to overcome [10].

Drought conditions lead to increased use of water resources in irrigated areas, but in non-irrigated fields obtaining necessary crop yields is a challenge. Corn, wheat, and barley grain yields have been shown to decrease as a result of drought [11–13]. Of importance to bioenergy technology developers planning to use lignocellulosic biomass, dry biomass yields of corn stover, switchgrass, and *Miscanthus* grown in research plots were reduced in the 2012 drought when compared to yields in 2011 and 2013 [14]. Even crops that have been reported to have some level of drought tolerance, like sorghum and switchgrass, had significant yield reductions during drought, 40–80% in some cases, even though the plants often survive the drought stress [15–17].

Drought is a major risk for producers and biorefineries relying on consistent and high crop yields; however, for the renewable energy industry the effect of drought on crops can be even more substantial and complex. The objective of this chapter is to discuss how biomass destined for renewable energy is affected by drought as it relates to overall dry biomass yields and chemistry, the latter of which heavily impacts cost of production and final product quality. The chapter proceeds with a discussion of how drought related risks impact the supply chain and strategies for risk reduction through thoughtful design of logistics systems for biorefineries. Finally, the chemical analysis of a variety of bioenergy crops grown during severe drought conditions as part of a set of long-term nationwide field trials will be discussed along with the state of knowledge regarding how these changes impact conversion to biofuels and products.

2. Reduction of bioenergy industry risk through supply system design

In bioenergy, the risks are as diverse as the economic agents that make up the industry. From the beginning of the supply chain, the risks farmers face are different from the risks aggregators face. Aggregators, the people who harvest, collect, and transport feedstocks to the biorefinery, face different risks than owners and managers of the biorefinery. While some risks in bioenergy apply across these agents, e.g., the risk that a market for the finished product might not exist, the fact that risk is perspective dependent means that one must be precise about whose risks are under discussion. This section considers one type of risk, supply risk, which biorefinery owners and managers face because of the role that drought and weather variability play.

Risk is a concept to measure 'unwanted' events. At the biorefinery, supply risk means that management must engage in unwanted, costly activity if the chance of insufficient feedstock supply delivered to the biorefinery for conversion materializes; thus the plant cannot run at full capacity. Risk is the probability of an unwanted event occurring multiplied by the consequence [18]. For management, this means that if the feedstock supply is lower than the full capacity of the plant then at least two undesirable events are realized [19, 20]. First, the amount of product created at the plant is reduced meaning that the unit cost of production, and the price necessary to cover costs, increases. The plant must utilize the same amount of resources to run the facility at full capacity as when it runs at less than full capacity, thus driving up unit cost. The second is that, in order to overcome the first undesirable outcome, management must seek out additional sources of feedstock supply. Because the additional supply is not placed under the contract managers have with growers for the initial supply, making up the short fall, too, is costly [20].

Depending on the nature of the risk under analysis, different approaches for mitigation apply [21]. Supply risk is considered non-systematic or diversifiable risk because yield uncertainty is not correlated with risks in other parts of the economic system. For example, crop yield does not correlate with stock market performance but instead with climate variability. For a biorefinery manager, non-systematic risk means that diversification is a strategy to mitigate weather variability. To illustrate how diversifying the feedstock supply allows the manager to mitigate risk, this section proceeds as follows. First, a description of the biomass feedstock supply chain provides a picture of how risk enters the system through yield uncertainty. Then an example illustrates how alternative supply system design enables mitigating supply risk caused by weather events, such as drought.

2.1 Supply chain

Possible biomass supply system configurations are numerous, but are typically classified in two ways: non-distributed and distributed. The non-distributed supply systems, also termed conventional supply systems, have been the systems of choice for the pioneer, or first-built biorefineries [22]. Non-distributed supply systems tend to be vertically integrated with a specific user. This means the biorefinery manages the supply chain from the time the biomass leaves the field to the time it enters the gates of the biorefinery. The materials are delivered in a minimally processed state and the burden of controlling and mitigating feedstock variability is placed on the users at the biorefinery. Non-distributed systems are typically sited in areas with an abundance of easily accessible and low-cost resources. The location of the pioneer biorefineries, as expected, have all been developed in areas with concentrated supplies of biomass known as supply sheds. While supply chains developed using this design are relatively uncomplicated and inexpensive, the biorefineries are limited to a small draw radius, due to the expenses associated with transporting material in the available formats. The relatively small supply shed may impact the ability for the biorefinery operators to mitigate feedstock quality issues with the resources available and potentially not be able to meet resource demands if there is a catastrophic event within the supply shed [23, 24].

The alternative, distributed supply system, sometimes called an advanced supply system, is a series of processing depots or terminals that are used to concentrate material from a small geographic region, near the point of production, and prepare it for use at a single or multiple facilities. This model is similar to how grain elevators work, the grain from local fields is aggregated and sold into a larger market. And, similar to logistics in grain supply systems, the processing depots may be owned by parties other than biorefinery owners. However, instead of simply holding the material for sale, the depots produce a stable, tradable intermediate product, which can be sold in a variety of markets. For a biorefinery, the largest benefit of the distributed supply system is having access to a larger supply shed for material. Biomass quality (e.g., ash and moisture content) is highly variable both spatially and temporally [23]. Through sourcing the material from a series of depots, biorefinery operators are able to specify the desired quality attributes of the material, and the burden of delivering material within the specifications is borne by the owners and operators of the depot. Although the cost of distributed supply systems seems high compared to a non-distributed system, given the requirements for additional infrastructure and increased transportation, system-wide benefits may offset costs [19]. The next section illustrates this point with an example of risk mitigation.

2.2 Mitigating drought risk

Figure 1 illustrates both the distributed and non-distributed, stylized supply chain configurations situated on a map of the Midwest United States. The panel on the left shows the location of a biorefinery and 10 potential sites for biomass depots. Multiple processing depots represent the advanced (distributed) case. The black lines illustrate the supply shed radius, which is the geographic area from which biorefinery management collects feedstock. In the conventional (non-distributed) case, the supply radius is 50 miles and the supply shed consists of fields near to the biorefinery. The dotted, black line next to the biorefinery shows the 50 miles radius. Economically constrained by transportation costs, in the conventional case management must contract with growers in near proximity to the biorefinery. On the other hand, the wider, solid black line encompasses the network of depots in the advanced supply case. Because of preprocessing, the economic constraint pushes the supply radius out to 400 miles, thus significantly expanding the supply shed. This enables management to contract with growers at much greater distance. The heat-map shading shows differing levels of drought intensity; red and orange illustrate greater drought intensity and blue a lesser amount.

The Year-A, Year-B designation in the left and right panels, respectively, shows two possible weather outcomes, generated with historical data. In Year-A the map does not show adverse weather events for either supply shed but in Year-B it shows adverse weather in much of the supply shed for both cases. While in Year-A none of the growers in the 50 miles supply shed experience detrimental impacts to crop yield from weather, in Year-B the growers next to the biorefinery collectively face the same adverse weather. By contrast, and looking at the 400 mile supply shed, growers in the northeast of the supply shed do not experience the adverse weather of much of the rest of the supply shed. A simulation model is a useful, analytic tool to understand how weather variability under these two supply chain configurations affect supply risk at the biorefinery.

Suppose management of the biorefinery in **Figure 1** contracts with growers for residual corn stover to procure feedstock to run a biorefinery with nameplate

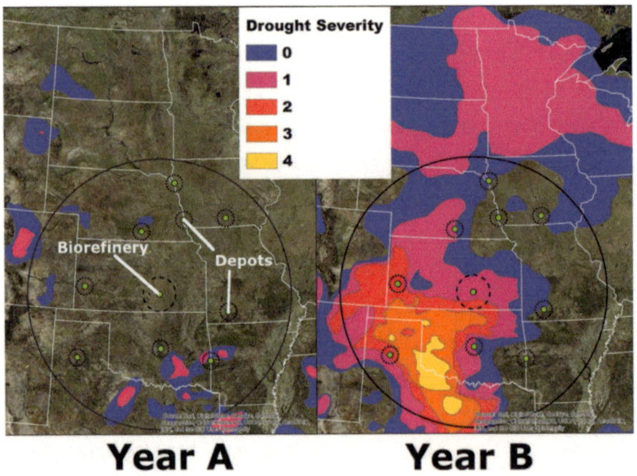

Figure 1.
Comparing two supply chain options under two weather scenarios based on historical data. Year A (left) has no adverse weather events, while year B (right) has moderate to severe drought covering much of the supply shed for both supply chain options [25].

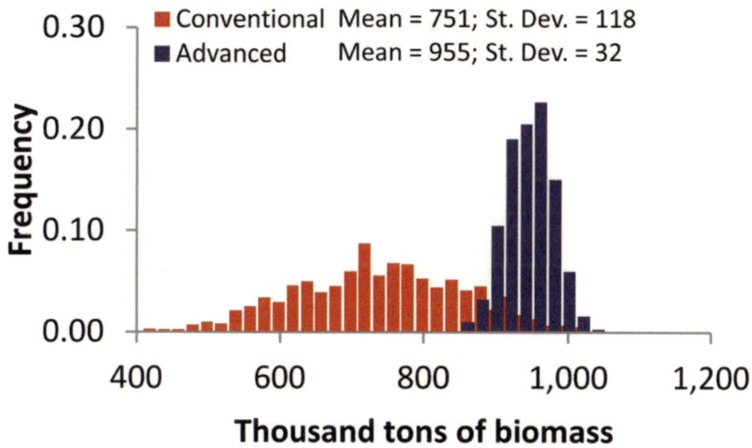

Figure 2.
Histogram of simulated feedstock outcomes under two supply chain configurations [25].

capacity of 800 thousand tons per year. In the conventional case and in the advanced case, management contracts with the same number of growers. In the conventional case farmers face the same distribution of yield uncertainty. In the advanced case the 10 distributions of uncertainty represent 10 separate regions of the supply shed.

Figure 2 shows the histogram of potential outcomes that result from a Monte Carlo simulation of the manager's contract options. The simulation utilizes parameters for yield, ash content, and dry matter loss that are representative of corn stover in the Midwest. The conventional case shows that on average, the manager will receive 751 thousand tons of biomass at the plant, but the range of possibilities extends from as little as 400 thousand tons to just over 1 million tons. In the advanced case, the histogram shows that the manager could expect on average 955 tons of biomass with a range of 800 thousand tons up to 1 million tons.

The results in the histogram illustrate the potential for risk reduction available to the manager by diversifying the supply portfolio. Much like one diversifies a financial retirement portfolio to mitigate risk, advanced supply configurations enable the same strategy. Managers at the biorefinery can mitigate drought-induced supply risk by diversifying the biorefinery's supply portfolio across a larger supply shed.

3. Drought impacts on chemical composition and conversion

3.1 Biomass chemical composition

The biomass supply risks related to drought are substantial and unfortunately extend to biomass quality as well as overall yields as discussed above. Crop yields are often reduced during drought conditions as plants do not have the water needed for basic functions like maintaining cell turgor pressure and performing photosynthesis [26]. The impact of drought conditions on yield as well as plant biochemical functions is complex and different plant types, species, and genotypes may vary in their tolerance and responses to drought [27, 28]. Species like *Miscanthus* are reported to be more sensitive to water deficiencies [29] while crops like sorghum [16], reed canary grass [27], and switchgrass [27, 30] display some level of drought tolerance. In addition, plants use different survival strategies to deal with environmental stressors; for example, there is less carbohydrate hydrolysis in cool-season forbs than in cool-season grasses during osmotic stress that occurs when plants experience drought requiring plant cell walls to compensate for the external changes in water with solute concentrations [31]. Soluble sugar synthesis in plants has been shown to occur in response to water stress in order to increase osmotic potential with significant accumulation of soluble sugars measured in switchgrass exposed to drought conditions [32].

Complicating reductions in crop yields, plants experience compositional changes during drought; increased extractive components, including soluble sugars, and decreased structural sugars were reported for important potential bioenergy crops like switchgrass, *Miscanthus*, mixed grasses, and corn stover [26, 32–34]. Studies have even observed reduced lignification in some cases possibly resulting from decreased plant growth as well as changes in lignin component distribution in plant cells impacting cell wall degradability [33–35]. These compositional changes can greatly impact yield of bioenergy conversion products from these biomass resources. It should also be noted that if these decreases in lignocellulosic components are compounded with decreases in dry biomass yield the estimated product yield can be even further reduced in drought-stressed crops [33].

To demonstrate the effect of large-scale drought on plant composition, data collected through the Regional Feedstock Partnership (RFP) was analyzed [36]. The RFP completed long-term field trials beginning in 2008 for potential bioenergy crops grown across the U.S. unintentionally providing a unique snapshot of how drought could impact the bioenergy industry when nationwide drought covered 65% of the continental U.S. in 2012 during the field trials [36–38]. Four RFP crops—*Miscanthus*, mixed grasses, switchgrass, and energycane—were selected to examine the impact of drought on plant chemistry. Each crop field site, according the U.S. Drought Monitor [37], experienced a year with drought conditions and a control year with minimal to no drought (**Figure 3**). *Miscanthus*, mixed grasses, and switchgrass data were from 2010, non-drought control year, and 2012, a year with significant drought. *Miscanthus* was located in Saunders County, NE; switchgrass in Day County, SD; and mixed grasses in Ellis County, KS each grown under three nitrogen application levels with three to four replicates. In Tift County, GA, where the energycane field site was located, a drought occurred in 2011 as opposed to 2012, and the non-drought control year used was 2009 as shown in the insets in **Figure 3**. Five genotypes were each grown on three replicate plots for the energycane field site.

Across four crop types, multiple energycane genotypes, and a variety of nitrogen fertilizer treatments, it is clear that biomass from drought years had lower lignocellulosic components than non-drought years, depicted by the differences in glucan, xylan, and lignin greater than zero (**Figure 4a–c**). Glucan was as much as 10% lower for biomass produced during a drought year (**Figure 4a**), while lignin was up to 5.5% lower (**Figure 4c**) and xylan up to 3.5% lower (**Figure 4b**). These differences are hypothesized to result from less lignification during reduced plant growth and increased synthesis of soluble components that support osmoregulation, in favor of synthesis of lignocellulosic components as hypothesized in previous studies [17, 33, 39]. It should be noted that not all research plots included in the analysis had greater lignocellulosic components in a non-drought year (differences less than zero shown by dotted lines in **Figure 4**), which is probably a result of the complex agronomic and environmental factors that can simultaneously impact plant yield and composition. Previously reported results on similar RFP samples indicated that along with the year-to-year variability, including drought and non-drought years, agronomic factors of nitrogen treatment and genotype also significantly impacted biomass yields and sustainability measurements [36]. Future studies are necessary to examine the complexity of the combination of these factors using multivariate analysis techniques that include, but are not limited to, drought. In addition, compositional changes in response to drought in the

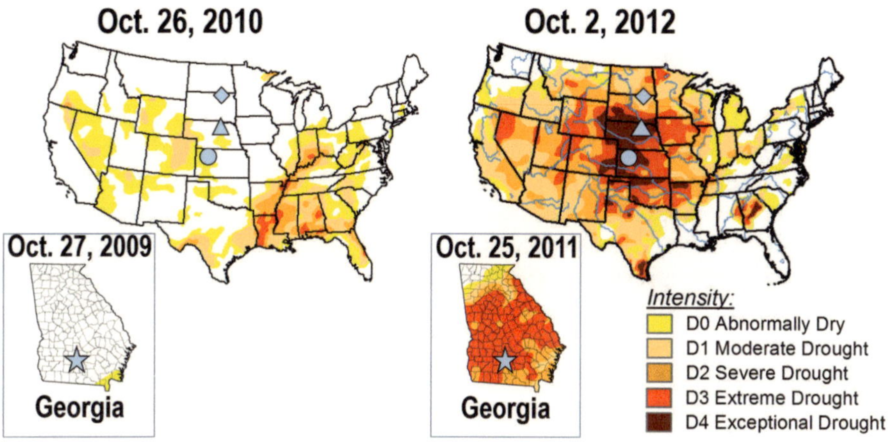

Figure 3.
U.S. Drought Monitor maps displaying continental U.S. drought conditions prior to the time of harvest in October 2010, a year with minimal drought, and 2012, a year with drought conditions covering much of the U.S. The inset displays Georgia drought conditions for October 2009 and 2011. Blue shapes indicate the approximate location of each biomass crop: triangles (▴) for Miscanthus, circles (•) for mixed grasses, diamonds (♦) for switchgrass, and stars (⋆) for energycane.

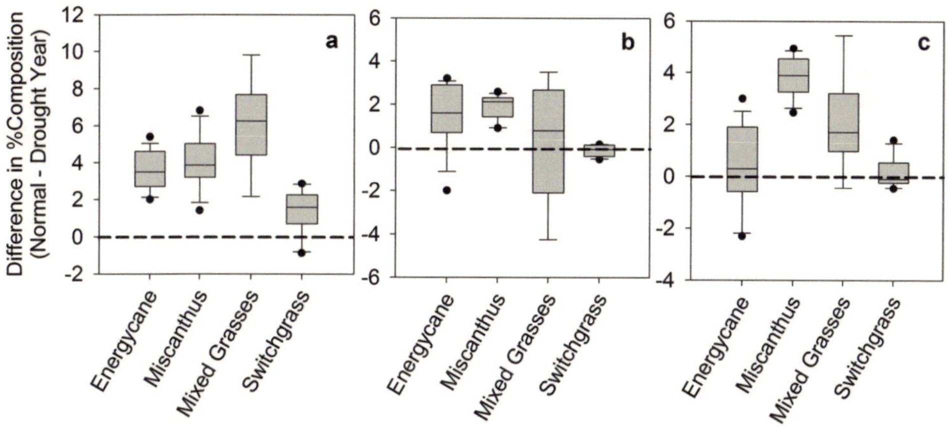

Figure 4.
*Difference between percent glucan (a), xylan (b), and lignin (c) content for four crops grown in a year with minimal to no drought and a year with drought conditions shown in **Figure 3**. Dotted lines represent no difference. Energycane—n = 15, Miscanthus—n = 12, mixed grasses—n = 9, and switchgrass—n = 11.*

literature are mixed. A number of studies report hemicellulose and lignin contents decreasing after drought treatments [33–35, 40], and in contrast other studies report hemicellulose and lignin contents remaining unchanged or increasing under drought conditions [33, 39, 41]. These differences are not completely understood; however, studies have suggested they arise from differences in drought severity and timing [33, 34, 39], genetics [35], and species specific differences [31, 42]. In addition, other environmental parameters like soil nutrient content and texture, timing of precipitation, growing degree days, and optimal growing temperatures also likely play a role.

3.2 Biochemical conversion processes

Drought-induced alterations to plant composition can significantly impact the yield of conversion products. Changes in biomass composition were exhibited by

RFP crops in response to drought stress, where the combined reduction of both structural carbohydrates and biomass yield led to an average 10–15% decrease in theoretical ethanol yield per Mg of dry biomass for *Miscanthus*, corn stover, and mixed perennial grasses [33]. In the 2012 drought year, mixed grasses grown in Kansas had only 10% of the dry biomass yield obtained in the non-drought year and *Miscanthus* dry biomass yield in Nebraska was reduced by an average of 14% [38]. These dry biomass decreases coupled with carbohydrate reductions shown in **Figure 4** severely reduce theoretical product yields. Interestingly, energycane in Georgia and switchgrass in South Dakota did not have dramatic decreases in above-ground biomass yield, which may be due to strong responses to other factors like temperature in the case of energycane, and the reported drought tolerance of switchgrass [38]. Theoretical ethanol yield is often used to demonstrate conversion potential for bioenergy crops based on carbohydrate compositions; however, it is just an estimate of potential yield and is based on assumptions of 100% conversion of carbohydrates to ethanol. In reality, there are many other considerations regarding biomass composition that can affect the pretreatment, enzymatic hydrolysis, and fermentation steps that are necessary to convert biomass to products in biochemical conversion. Hoover et al. [34] reported that *Miscanthus* carbohydrate yields from dilute-acid pretreatment and enzymatic hydrolysis were actually higher in drought affected plants compared to those grown in a non-drought year, which was hypothesized to be a result of higher extractable glucose and lower lignin contents. It is thought that reduced lignin content, observed in some drought-stressed plants, can decrease recalcitrance by creating better access to cell wall carbohydrates and increasing conversion efficiency, but changes in lignin distribution in tissues may also play a role in cell wall degradability in water stressed plants [35, 43]. The increase in carbohydrate yields is not isolated to dilute-acid pretreatment and enzymatic hydrolysis, as drought-stressed *Miscanthus* had increased carbohydrate yields in mild-alkali pretreatment and enzymatic saccharification [39] and after a mild hot water pretreatment and saccharification in nutrient rich environments [28]; in both studies this trend was either less pronounced or not present for leaves when compared to stems. A tall fescue mixture also had few significant increases in carbohydrate conversion yields, thought to be a result of less severe drought growing conditions [34]. A recent report documented increased extractability of pectin components in the cell wall ultrastructure of loblolly pine in response to low soil moisture [44]. Increases in cell wall elasticity have been observed under moisture stress conditions in *Pinus radiata* and may be related to drought tolerance [45]. Pattathil et al. [44] suggested that stress-induced alterations in cell wall elasticity may involve cell wall loosening processes that result from rearrangement of structural cell wall components like pectins and hemicelluloses. Increased elasticity of plant cell walls in biomass may pose further challenges to feeding, handling, and physical/mechanical deconstruction of biomass that is requisite for biochemical conversion. Understanding the changes in cell wall structure, chemical components, and physical properties imparted by drought stress is critical to informing how these properties can be exploited to improve bioprocessing of lignocellulosic feedstocks to biofuels and co-products.

It must also be considered how drought impacts the formation of certain degradation products that decrease conversion efficiencies though inhibition of enzymes during enzymatic hydrolysis and microorganisms during the fermentation step in a biochemical conversion process. For *Miscanthus* pretreated with dilute acid, enzymatic and fermentation inhibitors did not increase, however, this was likely a result of the dilute-acid pretreatment temperatures being lower than those required to form inhibitors [34]. In contrast, fermentation inhibitors were increased in drought stressed switchgrass in a study by Ong et al. [32] where the switchgrass was chemically pretreated using ammonia fiber expansion (AFEX). Specifically, the increase in soluble sugars formed during drought conditions generated pyrazines and imidazoles in the AFEX pretreatment that inhibited growth of the fermentation organism *S. cerevisiae*. These two studies highlight the need to understand drought impacts on plant biochemistry as well as intermediate and final product yields in order to mitigate these impacts. For example, technology developers in areas that are non-irrigated and prone to drought, either currently or in future climate scenarios, should consider the best pretreatment options for a biochemical conversion process based on the chemical profile of drought-stressed feedstock. In addition, a refiner could tailor pretreatment severity to the chemical composition of the biomass entering a facility to optimize product yields by limiting inhibitor formation with the least reduction in carbohydrate product yields. Future research and development activities might focus on enzyme and microorganism development to better handle inhibitors formed as a result of increased extractive components during uncontrollable environmental conditions. Finally, blending either prior to pretreatment or between different steps in the conversion process could be used to control intermediate or product yields and/or reduce concentrations of inhibitors to tolerable levels [46].

4. Conclusions

Drought is a risk for the bioenergy industry that is likely to increase in future years. Current knowledge and resources regarding drought impacts on crop yields, quality of biomass, and conversion performance can be used for determining research and development directions and mitigation strategies. Weather patterns and water resources are important considerations early in the process of site and feedstock selection for a facility where matching genotypes to conditions can support optimization of yields. Irrigation may be an option in certain cases, but there are implementation costs and water resources may not be an available or sustainable option given that a vast amount of water resources are currently consumed for agriculture. The scenarios in this chapter examine an alternative approach demonstrating that supply system design can reduce supply chain risk related to drought; these advanced supply systems hold promise for future biorefineries. Supply risk associated with drought needs to consider crop yield losses, in addition to biomass chemical changes. Data from a RFP field study of four energy crops, representing a variety of nitrogen application treatments and genotypes, showed how biomass lignocellulosic components—glucan, xylan, and lignin—were lower for a drought year compared to a non-drought year. Current literature was used to describe how drought related chemical changes propagate from the field through the conversion process, and planning and mitigation can be implemented throughout the system to reduce risk to the biomass producer and biorefinery. Drought induced chemical changes can create inhibitors during pretreatment, a step in biochemical conversion processes, that decrease the efficiency of the conversion process, which reinforces the need for careful selection of pretreatment methodology and severity based on location and biomass used. In addition, research and development is necessary for enzyme and microorganism development as well as to fully understand species' specific response to drought and support breeding programs to produce bioenergy cultivars with traits like increased water use efficiency. Finally, an advanced supply system can supply a refinery with more consistent biomass amounts year to year reducing operating risk, but a refinery may still receive feedstock with varying quality, even in a given year. Therefore, in-line techniques to monitor biomass chemistry entering a facility could be used to blend biomass or intermediates to specifications or adjust pretreatment severity to minimize degradation of soluble components generated during drought stress.

Acknowledgements

The authors would like to thank the Regional Feedstock Partnership (RFP) members, particularly those from South Dakota State University, Kansas State University, University of Georgia, and University of Nebraska, where the RFP experiments were executed and crops produced that were used in the analysis in this chapter. Drought maps in **Figure 3** were courtesy of the National Drought Mitigation Center at the University of Nebraska-Lincoln, the United States Department of Agriculture, and the National Oceanic and Atmospheric Administration. Authors of the continental U.S. maps are: Eric Luebehusen, USDA (2010); Anthony Artusa, NOAA/ NWS/ NCEP/CPC (2012). Authors of the Georgia maps are: Matthew Rosencrans, CPC/ NCEP/NWS/NOAA (2009); David Miskus, NOAA/NWS/NCEP/CPC (2011). Finally, the authors would like to thank the National Renewable Energy Laboratory for biomass characterization and the following Idaho National Laboratory colleagues: Garold Gresham, Leilani Beard, Mary Bingham, Karen Delezene-Briggs, Matthew Bryant, Sergio Hernandez, Sabrina Morgan, and Brad Thomas.

Financial and competing interests disclosure

This research was supported by the U.S. Department of Energy (DOE), Office of Energy Efficiency and Renewable Energy (EERE), Bioenergy Technologies Office (BETO), under DOE Idaho Operations Office Contract DE-AC07-05ID14517. The views expressed in the article do not necessarily represent the views of the U.S. Department of Energy or the United States Government. The US Government retains and the publisher, by accepting the article for publication, acknowledges that the US Government retains a nonexclusive, paid-up, irrevocable, worldwide license to publish or reproduce the published form of this manuscript, or allow others to do so, for US Government purposes. The authors have no other relevant affiliations or financial involvement with any organization or entity with a financial interest in or financial conflict with the subject matter or materials discussed in the manuscript apart from those disclosed. No writing assistance was utilized in the production of this manuscript.

Notes

Chemical composition data presented in this book chapter are available in the Bioenergy Feedstock Library (bioenergylibrary.inl.gov).

Author details

Amber Hoover*, Rachel Emerson, Jason Hansen, Damon Hartley and Allison Ray
Idaho National Laboratory, Idaho Falls, ID, USA

*Address all correspondence to: amber.hoover@inl.gov

IntechOpen

© 2019 The Author(s). Licensee IntechOpen. This chapter is distributed under the terms of the Creative Commons Attribution License (http://creativecommons.org/licenses/by/3.0), which permits unrestricted use, distribution, and reproduction in any medium, provided the original work is properly cited.

References

[1] Bauen A, Berndes G, Junginger M, Londo M, Vuille F, Ball R, et al. Bioenergy—A Sustainable and Reliable Energy Source: Main Report. IEA Bioenergy; 2009

[2] U.S. Department of Energy. In: Efroymson RA, Langholtz MH, Johnson KE, Stokes BJ, editors. Billion-Ton Report: Advancing Domestic Resources for a Thriving Bioeconomy, Volume 2: Environmental Sustainability Effects of Select Scenarios from Volume 1. Oak Ridge, TN, Oak Ridge National Laboratory, 2017 ORNL/TM-2016/727; 2016. DOI: 10.2172/1338837

[3] Stone KC, Hunt PG, Cantrell KB, Ro KS. The potential impacts of biomass feedstock production on water resource availability. Bioresource Technology. 2010;**101**(6):2014-2025. DOI: 10.1016/j.biortech.2009.10.037

[4] Peterson TC, Heim RR Jr, Hirsch R, Kaiser DP, Brooks H, Diffenbaugh NS, et al. Monitoring and understanding changes in heat waves, cold waves, floods, and droughts in the United States: State of knowledge. Bulletin of the American Meteorological Society. 2013;**94**(6):821-834. DOI: 10.1175/bams-d-12-00066.1

[5] Williams IN, Torn MS, Riley WJ, Wehner MF. Impacts of climate extremes on gross primary production under global warming. Environmental Research Letters. 2014;**9**(9):1-10. DOI: 10.1088/1748-9326/9/9/094011

[6] Wehner M, Easterling DR, Lawrimore JH, Heim RR Jr, Vose RS, Santer BD. Projections of future drought in the continental United States and Mexico. Journal of Hydrometeorology. 2011;**12**(6):1359-1377. DOI: 10.1175/2011jhm1351.1

[7] Hansen J, Sato M, Ruedy R. Perception of climate change. Proceedings of the National Academy of Sciences of the United States of America. 2012;**109**(37):E2415-E2423. DOI: 10.1073/pnas.1205276109

[8] Langholtz M, Webb E, Preston BL, Turhollow A, Breuer N, Eaton L, et al. Climate risk management for the US cellulosic biofuels supply chain. Climate Risk Management. 2014;**3**:96-115

[9] Rippey BR. The U.S. drought of 2012. Weather and Climate Extremes. 2015;**10**(Part A):57-64. DOI: 10.1016/j.wace.2015.10.004

[10] U.S. Department of Energy. Biorefinery Optimization Workshop Summary Report. 2016

[11] Earl HJ, Davis RF. Effect of drought stress on leaf and whole canopy radiation use efficiency and yield of maize. Agronomy Journal. 2003;**95**(3):688-696

[12] Samarah NH. Effects of drought stress on growth and yield of barley. Agronomy for Sustainable Development. 2005;**25**(1):145-149. DOI: 10.1051/agro:2004064

[13] Keyvan S. The effects of drought stress on yield, relative water content, proline, soluble carbohydrates and chlorophyll of bread wheat cultivars. Journal of Animal & Plant Sciences. 2010;**8**:1051-1060

[14] Sanford GR, Oates LG, Jasrotia P, Thelen KD, Robertson GP, Jackson RD. Comparative productivity of alternative cellulosic bioenergy cropping systems in the North Central USA. Agriculture, Ecosystems and Environment. 2016;**216**:344-355

[15] Barney JN, Mann JJ, Kyser GB, Blumwald E, Van Deynze A, DiTomaso JM. Tolerance of switchgrass to extreme soil moisture stress: Ecological

implications. Plant Science. 2009;**177**(6):724-732. DOI: 10.1016/j.plantsci.2009.09.003

[16] Gill JR, Burks PS, Staggenborg SA, Odvody GN, Heiniger RW, Macoon B, et al. Yield results and stability analysis from the sorghum regional biomass feedstock trial. Bioenergy Research. 2014;**7**(3):1026-1034. DOI: 10.1007/s12155-014-9445-5

[17] Perrier L, Rouan L, Jaffuel S, Clement-Vidal A, Roques S, Soutiras A, et al. Plasticity of sorghum stem biomass accumulation in response to water deficit: A multiscale analysis from internode tissue to plant level. Frontiers in Plant Science. 2017;**8**:1-14. DOI: 10.3389/fpls.2017.01516

[18] Kaplan S, Garrick BJ. On the quantitative definition of risk. Risk Analysis. 1981;**1**(1):11-27

[19] Lamers P, Roni MS, Tumuluru JS, Jacobson JJ, Cafferty KG, Hansen JK, et al. Techno-economic analysis of decentralized biomass processing depots. Bioresource Technology. 2015;**194**:205-213. DOI: 10.1016/j.biortech.2015.07.009

[20] Hansen JK, Jacobson JJ, Roni MS. Quantifying supply risk at a cellulosic biorefinery. In: Chichadly K, Saeed K, editors. 33rd International Conference of the System Dynamics Society; 19-23 July 2015. Cambridge, MA, USA: System Dynamics Society; 2015. pp. 1255-1279

[21] Hansen JK, Lipow J. Accounting for systematic risk in benefit-cost analysis: A practical approach. Journal of Benefit-Cost Analysis. 2013;**4**(3):361-373. DOI: 10.1515/jbca-2013-0008

[22] Hess J, Kenney K, Ovard L, Searcy E, Wright C. Commodity-Scale Production of an Infrastructure-Compatible Bulk Solid form Herbaceous Lignocellulosic Biomass. Idaho Falls, ID: Idaho National Laboratory; 2009

[23] Kenney KL, Smith WA, Gresham GL, Westover TL. Understanding biomass feedstock variability. Biofuels. 2013;**4**(1):111-127

[24] Argo AM, Tan EC, Inman D, Langholtz MH, Eaton LM, Jacobson JJ, et al. Investigation of biochemical biorefinery sizing and environmental sustainability impacts for conventional bale system and advanced uniform biomass logistics designs. Biofuels, Bioproducts and Biorefining. 2013;**7**(3):282-302

[25] Searcy E, Lamers P, Hansen JK, Jacobson J, Hess R, Webb E. Advanced Feedstock Supply System Validation Workshop Summary Report. 2015

[26] Chaves MM, Maroco JP, Pereira JS. Understanding plant responses to drought—From genes to the whole plant. Functional Plant Biology. 2003;**30**(3):239-264. DOI: 10.1071/fp02076

[27] Lewandowski I, Scurlock JMO, Lindvall E, Christou M. The development and current status of perennial rhizomatous grasses as energy crops in the US and Europe. Biomass and Bioenergy. 2003;**25**(4):335-361. DOI: 10.1016/s0961-9534(03)00030-8

[28] da Costa RMF, Simister R, Roberts LA, Timms-Taravella E, Cambler AB, Corke FMK, et al. Nutrient and drought stress: Implications for phenology and biomass quality in *Miscanthus*. Annals of Botany. 2018:1-14. DOI: 10.1093/aob/mcy155

[29] Heaton E, Voigt T, Long SP. A quantitative review comparing the yields of two candidate C-4 perennial biomass crops in relation to nitrogen, temperature and water. Biomass and Bioenergy. 2004;**27**(1):21-30. DOI: 10.1016/j.biombioe.2003.10.005

[30] Wright L, Turhollow A. Switchgrass selection as a "model" bioenergy crop: A history of the process. Biomass and

Bioenergy. 2010;**34**(6):851-868. DOI: 10.1016/j.biombioe.2010.01.030

[31] Karsten H, MacAdam JW. Effect of drought on growth, carbohydrates, and soil water use by perennial ryegrass, tall fescue, and white clover. Crop Science. 2001;**41**(1):156-166

[32] Ong RG, Higbee A, Bottoms S, Dickinson Q, Xie D, Smith SA, et al. Inhibition of microbial biofuel production in drought-stressed switchgrass hydrolysate. Biotechnology for Biofuels. 2016;**9**(1):237-250

[33] Emerson R, Hoover A, Ray A, Lacey J, Cortez M, Payne C, et al. Drought effects on composition and yield for corn stover, mixed grasses, and *Miscanthus* as bioenergy feedstocks. Biofuels. 2014;**5**(3):17

[34] Hoover A, Emerson R, Ray A, Stevens D, Morgan S, Cortez M, et al. Impact of drought on chemical composition and sugar yields from dilute-acid pretreatment and enzymatic hydrolysis of *Miscanthus*, a tall fescue mixture, and switchgrass. Frontiers in Energy Research. 2018;**6**:1-15. DOI: 10.3389/fenrg.2018.00054

[35] El Hage F, Legland D, Borrega N, Jacquemot MP, Griveau Y, Coursol S, et al. Tissue lignification, cell wall p-coumaroylation and degradability of maize stems depend on water status. Journal of Agricultural and Food Chemistry. 2018;**66**(19):4800-4808. DOI: 10.1021/acs.jafc.7b05755

[36] Owens VN, Karlen DL, Lacey JA, et al. Regional Feedstock Partnership Report: Enabling the Billion-Ton Vision. U.S. Department of Energy and Idaho National Laboratory; 2016. Document No.: INL/EXT-15-37477

[37] US Drought Monitor. National Drought Mitigation Center, University Nebraska-Lincoln [Internet]. Available from: https://droughtmonitor.unl.edu/Maps.aspx [Accessed: 22 January 2019]

[38] Owens VN. Sun Grant/DOE Regional Feedstock Partnership Final Technical Report. North Central Regional Sun Grant Center, South Dakota State University; 2018. Document No.: DOE-SDSU-85041

[39] Van Der Weijde T, Huxley LM, Hawkins S, Sembiring EH, Farrar K, Dolstra O, et al. Impact of drought stress on growth and quality of *Miscanthus* for biofuel production. GCB Bioenergy. 2017;**9**(4):770-782

[40] Vincent D, Lapierre C, Pollet B, Cornic G, Negroni L, Zivy M. Water deficits affect caffeate O-methyltransferase, lignification, and related enzymes in maize leaves. A proteomic investigation. Plant Physiology. 2005;**137**(3):949-960

[41] Ottaiano L, Di Mola I, Impagliazzo A, Cozzolino E, Masucci F, Mori M, et al. Yields and quality of biomasses and grain in *Cynara cardunculus* L. grown in southern Italy, as affected by genotype and environmental conditions. Italian Journal of Agronomy. 2017;**12**(4): 375-382. DOI: 10.4081/ija.2017.954

[42] Lu ZJ, Neumann PM. Water-stressed maize, barley and rice seedlings show species diversity in mechanisms of leaf growth inhibition. Journal of Experimental Botany. 1998;**49**(329):1945-1952. DOI: 10.1093/jexbot/49.329.1945

[43] Davin LB, Patten AM, Jourdes M, Lewis NG. Lignins: A twenty-first century challenge. In: Himmel ME, editor. Biomass Recalcitrance: Deconstructing the Plant Cell Wall for Bioenergy. Chichester, UK: Blackwell Publishing Ltd.; 2008. pp. 213-305

[44] Pattathil S, Ingwers MW, Victoriano OL, Kandemkavil S, McGuire MA, Teskey RO, et al. Cell wall ultrastructure of stem wood, roots, and needles of a conifer varies in response to moisture availability. Frontiers in Plant Science. 2016;7:1-11

[45] De Diego N, Sampedro MC, Barrio RJ, Saiz-Fernandez I, Moncalean P, Lacuesta M. Solute accumulation and elastic modulus changes in six radiata pine breeds exposed to drought. Tree Physiology. 2013;**33**(1):69-80. DOI: 10.1093/treephys/tps125

[46] Ray AE, Li C, Thompson VS, Daubaras DL, Nagle NJ, Hartley DS. Biomass blending and densification: Impacts on feedstock supply and biochemical conversion performance. In: Tumuluru JS, editor. Biomass Volume Estimation and Valorization for Energy. Rijeka, Croatia: InTech; 2017. pp. 341-359. DOI: 10.5772/67207

Chapter 4

Native Plants to Arid Areas: A Genetic Reservoir for Drought-Tolerant Crops

Ricardo Trejo-Calzada, Aurelio Pedroza-Sandoval, Jesus G. Arreola-Avila and Fabian García-González

Abstract

Droughts are common in arid areas. These cause important losses in crop production, while the increasing population demands more food and goods. Cultivars able to produce under drought conditions are required to avoid or reduce production losses. Plants have evolved different mechanisms to face drought, and many genes have been already discovered in model and cultivated plants that are involved in this trait. Some of these genes have been successfully transformed into cultivated plants for drought tolerance. Plants native to arid lands may possess variants of drought tolerance mechanisms as compared to mesophytic or model plants. Also, different drought-related genes can be revealed. Studies using high-throughput and bioinformatic tools may allow to discover new genes and give new insights on the mechanisms involved in drought tolerance. However, still scarce studies in plants native to arid lands show that there are many drought-related genes that have not been already characterized and potentially they may be novel genes. These novel genes may be used to improve crops for drought tolerance. Therefore, more physiological, transcriptomic, proteomic, and metabolomic studies are needed on plants native to the deserts.

Keywords: abiotic stress, water stress, drought-related genes, genetic diversity, deserts, vegetation, drought-tolerance mechanisms, oxidative stress, osmotic adjustment, differentially expressed genes

1. Introduction

Arid lands are defined by the United Nations Environment Programme (UNEP) based on the ratio of average annual precipitation and potential evapotranspiration or aridity index (AI). Arid lands are those with an AI lower than 0.65 [1]. Arid lands are widely spread around the world. They include around 41% of the earth's land surface [2]. These areas include hyperarid (1 billion ha), arid, semiarid, and dry subhumid (5.1 billion ha) regions distributed across virtually all the continents [3]. Arid lands are often affected by droughts, which disturb natural and managed ecosystems and cause less biomass production, biodiversity loss, poverty, and insecurity [2, 4].

Droughts are one of the main environmental factors that prevent plants from reaching their full genetic potential and strongly reduce plant growth and

IntechOpen

productivity of native populations and in agricultural systems of arid and semiarid areas [5, 6]. Droughts are increasing in frequency and intensity on large regions of the world due to desertification processes. Desertification as a result of greater world's population, extensive agricultural practices, and the effects of the global climate changes has become a serious problem [7]. Droughts are not unusual when viewed from a geological or evolutionary perspective. Therefore, they represent a permanent, strong, and increasing factor for biological, ecological, agronomic, and economic processes [8].

Plants as sessile organisms are under strong selective pressure to adapt to their environment. Along the entire life cycle, plants are frequently subject to a combination of different biotic and abiotic stresses [9]. Drought is one of the major environmental factors that reduce the yield of crops by limiting water availability for plants' growth [10, 11]. Water deficit is one of the most important abiotic stresses that affect development and productivity of crop plants [12]. Water stress induces stomatal closure, restricts gas exchange, and reduces photosynthesis in plants [13]. Plants must cope with their environment through physiological acclimation and evolutionary adaptation. As a result, plants have evolved extraordinary mechanisms to perceive, respond, and survive some abiotic stresses and especially with water stress [8].

Most of the cultivated plants are very sensitive to drought and water stress [14, 15]. Thus, agricultural production is seriously limited by drought, while world population is continuously and increasingly demanding for food [16]. New cultivars that can keep or increase yields while using less water are needed in order to reduce the risk of grave yield losses and to reach the world food requirement [17, 18].

Some plant species have evolved morphological, physiological, and biochemical mechanisms that allow them to survive, grow, and reproduce in areas where drought is a common event. Plants may deal with water stress through escape, avoidance, and tolerance strategies [19, 20]. Escape is reached through plastic shifts in phenology in such a way that critical growth periods do not coincide with periods of water deficit. Avoidance includes adaptive responses to keep plant water status during drought. Tolerance strategies protect plant cells, tissues, and organs from water stress letting recovery after periods of water deficit and drought. These strategies are not mutually exclusive and may have different functions depending on species and duration, intensity, and timing of water stress [21, 22].

Native plants to arid lands must deal with extreme environmental factors such as high temperatures, high evaporation, low precipitation, salinity and high light intensity, and low soil moisture [23, 24]. These plants have evolved numerous mechanisms to survive drought. Some of those mechanisms have been described for model and cultivated plants [25]. However, wild populations frequently contain large pools of genetic and phenotypic diversity that can be useful for identifying new molecular strategies involved in drought tolerance. Detecting and understanding the roles of novel genes in drought tolerance may be the basis to improve or develop cultivars tolerant to drought [26, 27]. In this chapter we review some of the findings about drought-responsive genes in plants native to arid lands.

2. Drought tolerance genes

Plants' drought tolerance is a quantitative trait that is controlled by many genes with additive effects [28, 29]. In native plants to arid areas such as *Cynanchum komarovii* and *Prunus mongolica* Maxim, more than 3000 differentially expressed genes (DEG) under drought stress have been found [30, 31]. These genes and/or their products participate in at least two pathways involved in drought tolerance: ABA-dependent and ABA-independent pathways [25, 32]. According to their roles, these genes have been grouped into two groups: (i) genes that code for regulatory proteins involved in signal transduction (transcription factors, protein kinases, ABA biosynthesis) and (ii) genes that code for functional proteins such as water and ion channels, detoxification enzymes, protective proteins (late embryogenesis abundant (LEA) proteins, chaperones), proteins involved in osmolyte biosynthesis, and proteases [25, 33].

2.1 Regulatory proteins

Plants must deal with drought by activating a complex signaling network that produces a variety of physiological responses and defense systems [34, 35]. The signaling pathways include a group of stress sensors, cellular signal transduction, and transcriptionally regulated networks. In order to respond to water stress, it must be perceived by specific receptors which transmit the stress signal into the cell and may trigger a series of signal transduction steps. [36]. The signal transduction in ABA-dependent pathways increasing ABA concentrations is sensed by receptors such as GTG1, GTG2, Mg protoporphyrin IX chelatase H subunit (CHLH/ABAR), and pyrabactin resistance 1/PYR1-like/regulatory component of ABA response 1 (PYR/PYL/RCARs) [37]. Once ABA binds to its receptor, ABI1 (ABA insensitive 1) phosphatase activity is blocked causing autophosphorylation and activation of an open serine-threonine kinase (open stomata 1, OST1). Then, transcription factors (TFs) known as ABA-responsive element-binding proteins (ABFs/AREBPs) are phosphorylated and activated by OST1. Besides OST1, calcium-dependent kinases (CDPKs) similarly activate SLAC1 and SLAH3 [38]. Afterwards, ABFs bind to DNA at specific ABA-responsive elements (ABREs). Finally, ABA-dependent gene expression is induced [39].

Gene expression at the transcriptomic level is strongly regulated by transcription factors [40]. TFs can be grouped into several families according to their structure and binding domains [41]. Several TFs belonging to MYB, MYC, NAC, bZIP, HD-ZIP, DREB, and WRKY families are involved in modulation of gene expression of plants in response to drought stress through ABA-dependent or ABA-independent pathways [42]. Even though many TF families have been found in several models and cultivated plants [43, 44], novel TFs are being described from plants native to arid lands. They may be used as a powerful tool for practical approaches for engineering drought stress tolerance in plants. For example, in the desert legume *Eremosparton songoricum*, a novel DREB2B (EsDREB2B) gene was identified. The transcript of EsDREB2B was upregulated by different abiotic stresses, among them drought stress [45]. Other novel DREBs from a desiccation tolerant moss (*Syntrichia caninervis* Mitt.) were isolated and used to transform yeast. The ScDREB enhanced stress tolerance to yeast [46]. In *Larrea tridentata*, a bush native and widely distributed to Northern American deserts, the LtWRKY21 transcription factor that functions downstream of ABI1 to control ABA-dependent expression of genes was found [47]. The PeDREB2L gene from the desert tree *Populus euphratica* Oliva was isolated and transformed into *Arabidopsis thaliana*. Transgenic plants showed an improved tolerance to drought and freeze [48]. Moreover, in *Sophora moorcroftiana*, an endemic Leguminosae shrub species native to arid and semiarid regions of the Qinghai-Tibet Plateau, a total of 1534 TFs were identified. Those TFs were classified into 23 different common families. The major group of TFs was the bZIP family (160, 10.43%), followed by MYB (115, 7.5%), bHLH (107, 6.98%), zinc finger (103, 6.71%), and WRKY (103, 6.71%) [49].

2.2 Functional proteins

Drought stress affects many processes in the plants and cause a variety of physiological and biochemical changes. Some of these changes include loss of cellular turgor, changes in membrane fluidity and composition, changes in osmotic potential, and protein-protein interaction [10]. Cell turgor loss is perhaps the most evident indicator of water stress which affects integrity of cells, metabolism, and whole plant performance [50]. Maintaining cell turgor is critical for surviving and growth of plants. The changes in osmotic potential play a relevant role for that purpose. Loss of cell turgor, among other effects, may cause stomata closure and limitations of gas exchange which in turn decreases CO_2 supply for RuBisCo. Photosynthesis decreases, and the reducing power production exceeds the rate of its use by the Calvin cycle [51]. Consequently, overproduction and accumulation of reactive oxygen species (ROS) alter the redox status [52, 53]. ROS damage all major cell biomolecules impairing their function [53]. Plants may respond to these effects by activating several defense mechanisms that involve participation of numerous proteins such as late embryogenesis abundant (LEA) proteins, osmoprotectants, chaperons, detoxifying enzymes, and various proteases [25, 54].

2.2.1 Late embryogenesis abundant (LEA) proteins

A relatively well-known family of drought-responsive genes is the late embryogenesis abundant (LEA) gene family. LEAs are proteins that accumulate at late stage of development of many plant seeds. LEA proteins also accumulate in vegetative parts of plants as a response to water and osmotic stresses and ABA application [55]. It has been proposed that LEA proteins have an important role protecting cellular structures from water deficit [56]. LEA proteins have been grouped into at least eight different groups according to their amino acid sequence similarities and repeated sequence motifs (LEA1 to LEA6, dehydrins, and seed maturation protein). Most LEA proteins are highly hydrophilic, glycine-rich, and low-complexity proteins. They have a strongly disordered conformation in the hydrated state [57]. LEA genes are highly diverse and have been found in a wide range of plant species [58]. Up to 242 LEA genes have been found in *Gossypium hirsutum* [57]. The LEA genes have several stress-responsive cis-acting regulatory elements in the promoter region such as ABRE, DRE/CRT, MYBS, and LTRE [59]. The expression of LEA proteins is associated with acquisition of drought stress tolerance [60, 61]. The specific cellular role of LEA proteins is not well known. However, different studies have shown that LEA proteins may function in scavenging free radicals and ions; stabilization of enzymes, proteins, and membranes; interactions with RNA and DNA; and water retention during drought and other abiotic stresses [61].

Several novel LEA genes have been found in plants native to arid lands, and they have been successfully applied to transform plants for drought stress tolerance. For example, [62] cloned a LEA gene from *Tamarix androssowii*, a shrub that grows in arid or saline environments. The cloned LEA gene was transformed into tobacco. The transgenic plants showed less ion leakage and MDA content under drought than nontransgenic plants. Also, transgenic plants had a greater growth and lower number of wilted leaves. Also [63] use a *Tamarix androssowii* LEA gene (TaLEA) to obtain transgenic lines of *Populus simonii* × *P. nigra* which were compared to nontransgenic plants under salt and drought stress. They found that the constitutive expression of TaLEA in transgenic poplars improved salt and drought tolerance, which was attributed to the protection of cell membranes from damage. Moreover, [64] transformed a dehydrin protein from the desert grass *Cleistogenes songorica* (CsLEA) into alfalfa (*Medicago sativa* L.). Transgenic plants grew more than wild-type plants under drought stress. Moreover, transgenic plants were able to return to normal after rewatering. Identification of novel LEA genes in plants adapted to arid lands may be useful for improving drought tolerance of cultivated plants.

2.2.2 Osmoprotectants

Some plants respond to drought stress via organic and/or inorganic solute accumulation. Osmolytes or compatible solutes are produced in plants under drought or saline stress. They are small and neutral and do not affect plant metabolism even at relatively high concentrations [21]. These compounds are also called osmoprotectants and are included into three major chemical groups: amino acids (e.g., proline), polyols (e.g., mannitol, trehalose, fructans), and quaternary amines (e.g., glycine betaine and polyamines) [65]. Active accumulation of osmoprotectants may help plants to endure water stress by maintaining the fluidity of cell membranes, protecting and stabilizing proteins and macromolecular structures, detoxification of free radicals, and osmotic adjustment. Osmotic adjustment is a mechanism that allows plants to keep a flow of water to the cells. A greater concentration of osmolytes reduces the osmotic potential of cells, which in turn produces water movement to the cells, which allows maintenance of turgor. Osmotic adjustment also involves lowering of toxic concentrations of Na^+ by limiting influx, sequestration, or exclusion [66]. Therefore, cell activities take place at approximately normal speed for better growth and development of plants [67].

Several genes associated with synthesis of osmoprotectants have been identified, isolated, and clonated from a variety of plant species including those native to arid lands [68]. For example, the pyrroline-5-carboxylate synthase (P5CS) gene involved in proline synthesis was characterized in *Calotropis procera* from de novo assembled transcriptome contigs of a high-throughput sequencing dataset [69]. Also, [70] transformed wheat plants with a P5CS gene from *Vigna aconitifolia* a legume grown in arid and semiarid regions of India. They found that transgenic wheat acquired drought tolerance by proline accumulation, which may have protected plants against oxidative stress. A betaine aldehyde dehydrogenase (BADH) gene from *Atriplex canescens*, a perennial bush native to arid lands of Northern America, was introduced into a soybean cultivar. The expression of AcBADH increased after drought treatment of transgenic plants. Besides glycine betaine, proline content also increased, and transgenic soybean lines yielded up to 8.8% more than control plants under drought treatments [71]. A plasma membrane intrinsic protein (PIP) gene (ScPIP1) from *Simmondsia chinensis,* a typical desert shrub, was cloned and overexpressed in *Arabidopsis thaliana.* ScPIP1 conferred drought and salt tolerance probably by reducing membrane damage and increasing osmotic adjustment [72].

2.2.3 Detoxifying enzymes

Drought as other kinds of abiotic and biotic stresses may produce oxidative stress. The oxidative stress is caused by at least two processes: (i) an imbalance of production and detoxification of reactive oxygen species (ROS) and (ii) de novo ROS biosynthesis as a response for defense and adaptation to environment. These processes produce significant variations in the general cellular redox state [53, 73]. The term ROS includes substances with one or more activated atoms of oxygen that can be radicals. Some free radicals do not contain oxygen atoms. The key ROS are triplet oxygen, singlet oxygen, superoxide anion radical, hydrogen peroxide, and hydroxyl radical [53]. Oxidative stress causes lipid peroxidation that impairs membranes and induces loss of their barrier function, and consequently a breakdown of organelles occurs [74]. Plants have evolved defense systems against excess of ROS. Those systems include nonenzymatic and enzymatic responses. The enzymatic system consists of several enzymes such as superoxide dismutase (SOD), catalase (CAT), ascorbate peroxidase (APX), glutathione reductase (GR), and others that eliminate or scavenge ROS [75].

Plants native to arid lands may have antioxidant defense systems to limit the deleterious effects of ROS. For example, in *Oudneya africana*, a Saharan plant, water deficit caused variations in enzymatic and nonenzymatic antioxidants, differentially affecting the concentration of SOD and POX [76]. Transgenic poplar expressing the eukaryote translation initiation factor 5A (TaeIF5A1) from *Tamarix androssowii* showed greater superoxide dismutase (SOD) and peroxidase (POD) activities, lower electrolyte leakage, and improved tolerance to abiotic stresses [77]. Six plant species from semiarid Loess Hilly Region of China showed significant differences in SOD and POD activities when subject to drought along 3 months. Also there were significant interactions of SOD, CAT, and POD activities and MDA content between months and species [78]. A transcriptomic study under drought stress showed that *Prunus mongolica* Maxim, a species widely distributed in the Gobi Desert, increased transcription of iron superoxide dismutase and manganese superoxide dismutase which promoted drought stress tolerance [31].

3. Findings of novel genes in native plants from arid land areas

The development of high-throughput next-generation sequencing technologies offers opportunities for genome-wide transcription analysis of non-model plants and especially of desert plants and discovery of novel pathways and genes related to drought tolerance. These novel genes may be useful to improve drought tolerance in cultivated plants. Even though there are not many studies in plants native to arid regions, the results of some of the research that different groups have carried out allow insights over the abundance of potential novel genes for drought tolerance.

By using the analysis of subtracted expressed sequence tags (ESTs) in horse gram (*Macrotyloma uniflorum* (Lam.) Verdc.), 531 unigenes were found as upregulated by drought. Among these unigenes 366 showed significant similarity to known sequences in the database. Approximately 30% of the ESTs had no similarity to known proteins in the GenBank database. Those are considered as novel. Also, according to the functional classification, the most abundant ESTs were those related to stress responses (12%), DNA processing and nucleotide metabolism (10%), photosynthesis and electron transport (9%), and transcription factors (5%). There was redundancy in several known stress-responsive clones in dehydrated sample. The most prominent were metallothionein, glutathione S-transferase, RabGAP/TBC domain-containing protein, lipoxygenase, translationally controlled tumor protein, chaperon, lipid transfer protein, cysteine proteinase, calmodulin, calmodulin-binding protein, and sterol 24-C-methyltransferase [79]. Moreover, a protein-protein interactome study carried out in *M. uniflorum* showed that the highest number of PIPs occurred in shoot (416) and root (2228) tissues of a drought-tolerant genotype as compared to shoot (136) and root (579) tissues of a sensitive genotype. The PIPs most responsive to drought stress were kinase and transferase activities involved in signal transduction, cellular processes, nucleocytoplasmic transport, protein ubiquitination, and localization of molecules. These PIPs could be enclosed in mechanisms of drought tolerance of *M. uniflorum*. Also, they could provide new understandings of mechanisms involved in drought tolerance [80].

A transcriptomic analysis of the roots of *Ammopiptanthus mongolicus*, an endemic species to the Gobi Desert, allowed to identify 27 drought-responsive genes. These genes were grouped into the GO categories of response to osmotic stress, response to oxidative stress, response to hormone stimulus, and response to light stimulus. A total of 9771 (34%) out of 29,056 ESTs matched to known proteins recorded in the PlantGDB database. This means that many ESTs could be novel drought-responsive genes [81]. Another study of comprehensive transcriptome of *A. mongolicus* identified up to 6102 DEGs under drought stress at 3 points of exposure. A group of 2028 DEGs were common for the 3 points. Among them 779 DEGs were upregulated and 1185 were downregulated by drought. The upregulated DEGs included a heavy representation of genes encoding ripening-related proteins, LEA proteins, peroxidases, transporters, enzymes of flavonoid biosynthetic pathways, protein kinases, ethylene receptors, and transcription factors. About 17.2% of the common DEGs had no homology to known functional proteins [82]. Similarly, Pang et al. [83] identified 1620 DEGs, including 1106 upregulated DEGs and 514 downregulated DEGs.

In *Cynanchum komarovii*, a xerophytic plant species, up to 3134 unigenes were found as differentially expressed genes (DEGs) under drought stress. A total of 601 unigenes were induced, while 2533 unigenes were repressed. The most abundant upregulated DEGs were into the following GO groups: "oxidation-reduction process" with 120 DEGs, "single-organism metabolic process" with 185 DEGs, and "oxidoreductase activity" with 113 DEGs. The most downregulated genes were grouped into the following GO terms: "metabolic process," "cellular component organization or biogenesis," "cellular response to stimulus," "macromolecular complex," "cytoplasm,", "protein binding," and "hydrolase activity." KEGG pathway analysis allowed to identify that DEGs involved in "cutin, suberine, and wax biosynthesis" changed significantly, which may be due to the important role of this pathway for this plant species protection when exposed to drought [30].

Reaumuria soongorica is an extreme xerophyte shrub from Gobi and marginal loess of central Asia. A transcriptomic analysis identified 123 unigenes potentially associated with drought adaptation. A total of 46 unigenes were related to drought escape mechanisms, while 40 unigenes were potentially involved in drought avoidance. Also, 32 unigenes had identity to genes involved in ABA-dependent, while 8 unigenes had homology to some genes involved in ABA-independent pathways of drought tolerance. There was a 7.96% of unigenes that did not match any homologous genes in the known plant species [84]. Another study identified 1325 DEGs, including 379 upregulated DEGs and 946 downregulated DEGs under drought stress. Among DEGs, 20 genes encoded for kinases, and 14 encoded for transcription factors such as WRKY, NAC, MYC, TCP, and bZIP. Also 13 DEGs encoding for functional proteins such as LEA proteins, small heat shock proteins, and aquaporin (AQP) and proline transporter were identified. Moreover, 14 DEGs were found encoding for low-temperature-induced protein, dehydration-induced protein, defensing precursor, resistance protein, universal stress protein, and protein involved in protein kinase [85].

Haloxylon ammodendron (C.A.Mey) is a desert tree distributed in Middle and Central Asia. By doing a transcriptomic analysis in this species, up to 1060 unigenes were identified as DEGs for drought stress. Among them, 356 DEGs were upregulated and 704 DEGs were downregulated. A total of 469 (44%) of DEGs did not show homology to genes in NCBI database. Approximately 12.1% of DEGs with homology to known genes were associated with nitrogen metabolism, starch and sucrose metabolism, and fatty acid metabolism. Also, 35 DEGs encoded known or putative transcription factors such as WRKY, MYB, and ethylene-responsive [86].

The expression of drought-responsive genes may depend on the stress intensity. A study in *Sophora moorcroftiana* revealed that more genes were differentially expressed under severe water stress than mild stress. Up to 5648 unigenes were differentially expressed between control and severe stress plants. Around 601 unigenes were common for mild and severe water stress. Eleven out of 1534 TFs were selected for expression analysis. Among these, seven were drought-responsive. Those encoding for DREB, zinc-finger protein (ZnF), zinc-finger protein kinase (ZFPK), MYB, NAC, and WRKY were upregulated, while ERB was downregulated. Three selected aquaporins (AQPs) and one sugar transporter (SUT) genes were upregulated by drought. Genes encoding for scavenging reactive oxygen species such as POD, PRX, and GPDH were induced by drought stress [49].

Prunus mongolica is a plant native to the Gobi Desert. In this species a total of 3365 differentially expressed transcripts (DETs) for water stress were identified, counting 1876 transcripts upregulated and 1489 transcripts downregulated. Among these, 42 transcripts coding for 5 aquaporin subfamilies were found. Also, 15 potential plasma membrane intrinsic proteins (PIPs) were upregulated and 1 downregulated. Interestingly, 177 transcripts related to ROS scavenging were identified. Approximately 28% of them were predicted as SOD. Several significant pathways were identified to be related to drought tolerance of *P. mongolica*. These pathways included transcription factors and plant hormone signal transduction, starch and sucrose metabolism, and cysteine and methionine metabolism [31].

Artemisia sphaerocephala is a species found in sand dunes in the deserts of Northwest China. A transcriptomic analysis in this species identified 108 unigenes related to drought stress tolerance. These had homology to 17 kinases, 2 potential chaperones, 52 enzymes, 6 transporters and channels, and 3 aquaporins. Even though, transcription factor were not identified, 25 out of the 108 unigenes were among the 1000 most highly expressed genes [87].

Zygophyllum xanthoxylum is a succulent halophythe adapted to arid environments. In this plant a total of 1723 DEGs were identified as upregulated in leaves of plants under osmotic stress (−0.5 MPa) during 24 h. Up to 53 DEGs related to ROS scavenging were also identified as upregulated, most of them encoded for glutathione S-transferase (GST) and peroxidases (POD). Also, 31 DEGs homologous to transporters were upregulated in the roots under salt and osmotic stress. However, 23 DEGs related to photosynthesis were downregulated under osmotic stress [88].

Cleistogenes songorica is a C4 xerophyte widely distributed in the arid regions of Northwest China. A mining study for LEA genes identified at least 44 putative LEA proteins. They were named CsLEA1 to CsLEA44 and classified into eight subfamilies. These LEAs were characterized, and two unusual LEA stress-related domains, water stress and hypersensitive response (WHy), and LEA14-like desiccation-related proteins were detected. Most LEAs within the same family showed similar structures and properties. All the CsDehydrin proteins contained the YKS segments which are essential for keeping the capacity to adjust their conformation and maintain cellular homeostasis under stress conditions. Among the cis-regulatory elements of the LEA genes, more than three G-Boxes were registered for each *dehydrin*, *LEA_2*, and *SMP* genes and more than two MBS for *dehydrin* and *LEA_2* subfamily gene [89].

Prosopis juliflora is a species native to arid lands of Mexico and widely distributed in arid and semiarid regions of Central and South America. In this species 6874 DEGs were found under salt and drought stress. Approximately 42.6% (2932 DEGs) had homology to genes with GO annotation. Among these, 1339 DEGs were upregulated, while 1596 were downregulated. More DEGs under salt and drought stress were found in roots than those of leaves. Under drought stress, there was more upregulation than downregulation in leaves, while in roots there was more downregulation. A total of 30 unigenes were recognized as exclusively responsive to drought stress. One of these genes encoding for *Arabidopsis* ortholog ABR1 was downregulated. Two upregulated genes encoded for transcription factors involved in diminishing intracellular H_2O_2 levels. Also, three highly upregulated DEGs encoded for "pq-loop repeat family protein transmembrane family protein" which has been assigned with a potential role in stress tolerance. Many drought-responsive genes were tissue specific. For example, 1040 genes were specifically expressed in root tissues. Up to 805 genes were commonly regulated by drought and salt stress in root tissue. Upregulated genes included those coding for dehydration-responsive protein rd22, pectinesterase-2, LEA protein, and POD. Moreover, 74 DEGs were commonly upregulated in leaf tissues and 16 were downregulated. The induced genes included pectinesterase-2, non-specific lipid-transfer protein, delta-1-pyrroline-5-carboxylate synthase, and ras-related protein rabc2a [90].

4. Conclusions

Drought stress tolerance is a complex quantitative trait that involves several mechanisms and multiple genes to produce the defense responses of the plants for dealing with environmental conditions of drought. Studies in model plants have found some of the main mechanisms, genes, and proteins that are activated upon drought stress. However, not enough studies have been carried out in plants native to arid lands. The new sequencing technologies and bioinformatic tools applied to some plants native to desert areas have allowed to identify many of the already characterized genes and proteins involved in drought stress tolerance. The most important is that this kind of studies has shown that myriads of drought-related transcripts are not characterized and may belong to novel genes. These may be part of the already known pathways of plants responses to drought or may participate in novel ways to tolerate drought, and after further studies may be used to improve drought tolerance in crops.

Conflict of interest

The authors of this work declare no conflict of interest.

Author details

Ricardo Trejo-Calzada*, Aurelio Pedroza-Sandoval, Jesus G. Arreola-Avila and Fabian García-González
Chapingo Autonomous University, Bermejillo, Durango, México

*Address all correspondence to: rtrejo@chapingo.uruza.edu.mx

IntechOpen

© 2019 The Author(s). Licensee IntechOpen. This chapter is distributed under the terms of the Creative Commons Attribution License (http://creativecommons.org/licenses/by/3.0), which permits unrestricted use, distribution, and reproduction in any medium, provided the original work is properly cited.

References

[1] United Nations Environment Programme (UNEP). World Atlas of Desertification. London: Edward Arnold; 1992

[2] Food and Agriculture Organization (FAO). Global Guidelines for the Restoration of Degraded Forests and Landscapes in Drylands: Building Resilience and Benefiting Livelihoods. Forestry Paper No. 175. Rome: Food and Agriculture Organization of the United Nations; 2015

[3] Middleton N, Thomas D. World Atlas of Desertification. London: John Wiley & Sons, Inc; 1997

[4] MEA. Chapter 22: Millennium ecosystem assessment: Ecosystems and human well-being: Current state and trends. In: Bartlett Millennium Ecosytem Assessment (MEA). Vol. 1. Washington, DC: World Resources Institute; 2005

[5] Zhu JK. Salt and drought stress signal transduction in plants. Annual Review of Plant Biology. 2002;**53**:247-273

[6] Martínez JP, Ledent JF, Jajji M, Kinet JM, Lutts S. Effect of water stress on growth, Na^+ and K^+ accumulation and water use efficiency in relation to osmotic adjustment in two populations of *Atriplex halimus* L. Plant Growth Regulation. 2003;**41**:63-73

[7] Dai A. Increasing drought under global warming in observations and models. Nature Climate Change. 2013;**3**:52-58

[8] Juenger TE. Natural variation and genetic constraints on drought tolerance. Current Opinion in Plant Biology. 2013;**16**:274-281. DOI: 10.1016/j.pbi.2013.02.001

[9] Tuskan GA, Difazio S, Jansson S, et al. Genome of black cottonwood, *Populus trichocarpa* (Torr. & Gray). Science. 2006;**313**:1596-1604

[10] Chaves MM, Maroco JP, Pereira JS. Understanding plant responses to drought—From genes to the whole plant. Functional Plant Biology. 2003;**30**(3):239-264

[11] Salekdeh GH, Reynolds M, Bennett J, Boyer J. Conceptual framework for drought phenotyping during molecular breeding. Trends in Plant Science. 2009;**14**:488-496

[12] Okcu G, Kaya MD, Atak M. Effects of salt and drought stresses on germination and seedling growth of pea (*Pisumsativum* L.). Turkish Journal of Agriculture and Forestry. 2005;**29**:237-242

[13] Lawson T, Oxborough K, Morison JIL, Baker NR. The responses of guard and mesophyll cell photosynthesis to CO_2, O_2, light, and water stress in a range of species are similar. Journal of Experimental Botany. 2003;**54**:1743-1752

[14] Tuberosa R, Salvi S. Genomics-based approaches to improve drought tolerance of crops. Trends in Plant Science. 2006;**11**:405-412

[15] Shabala S. Learning from halophytes: Physiological basis and strategies to improve abiotic stress tolerance in crops. Annual Reviews. 2013;**112**:1209-1221

[16] Mir RR, Zaman-Allah M, Sreenivasulu N, Trethowan R, Varshney RK. Integrated genomics, physiology and breeding approaches for improving drought tolerance in crops. Theoretical and Applied Genetics. 2012;**125**:625-645

[17] Barnabas B, Jager K, Feher A. The effect of drought and heat stress on reproductive processes in cereals. Plant, Cell & Environment. 2008;**31**:11-38

[18] Fleury D, Jefferies S, Kuchel H, Langridge P. Genetic and genomic tools to improve drought tolerance in wheat. Journal of Experimental Botany. 2010;**61**:3211-3222

[19] Levitt J. Response of Plants to Environmental Stresses. New York: Academic Press; 1972

[20] Ludlow MM. Strategies of response to water stress. In: Kreeb KH, HRTM H, editors. Structural and Functional Responses to Environmental Stresses. The Hague: SPB Academic; 1989. pp. 269-281

[21] Bartels D, Sunkar R. Drought and salt tolerance in plants. Critical Reviews in Plant Sciences. 2005;**24**:23-58

[22] Chaves MM. Oliveira: Mechanisms underlying plant resilience to water deficits: Prospects for water-saving agriculture. Journal of Experimental Botany. 2004;**55**:2365-2384

[23] McNeely JA. Biodiversity in arid regions: Values and perceptions. Journal of Arid Environments. 2003;**4**(1):61-70

[24] Yates SA, Chernukhin I, Alvarez-Fernandez R, Bechtold U, Baeshen M, Baeshen N, et al. The temporal foliar transcriptome of the perennial C3 desert plant *Rhazya stricta* in its natural environment. BMC Plant Biology. 2014;**14**:2

[25] Shinozaki K, Yamaguchi-Shinozaki K. Gene networks involved in drought stress response and tolerance. Journal of Experimental Botany. 2007;**58**:221-227

[26] Kooyers NJ. The evolution of drought escape and avoidance in natural herbaceous populations. Plant Science. 2015;**234**:155-162

[27] Zhan L, Shu H, Zhang AY, Liu BL, Xing GF, Xue JA, et al. Foxtail millet WRKY genes and drought stress. Journal of Agricultural Science. 2017;**155**:777-790. DOI: 10.1017/S0021859616000873

[28] Mohammadi M, Taleei A, Zeinali H, Naghavi MR, Ceccareli S, Baum G. M: QTL analysis for phenologic traits in doubled haploid populations of barley. International Journal of Agriculture and Biology. 2005;**5**:820-823

[29] Fan XD, Wang JQ, Yang N, Dong YY, Liu L, Wang FW, et al. Gene expression profiling of soybean leaves and roots under salt, saline-alkali and drought stress by high-throughput Illumina sequencing. Gene. 2013;**512**:392-402

[30] Ma X, Wang P, Zhou S, Sun Y, Liu N, Li X, et al. De novo transcriptome sequencing and comprehensive analysis of the drought-responsive genes in the desert plant *Cynanchum komarovii*. BMC Genomics. 2015;**16**:753. DOI: 10.1186/s12864-015-1873-x

[31] Wang J, Zheng R, Bai S, Gao X, Liu M, Yan W. Mongolian almond (*Prunus mongolica* maxim): The morpho-physiological, biochemical and transcriptomic response to drought stress. PLoS One. 2015;**10**(4):e0124442. DOI: 10.1371/journal.pone.0124442

[32] Xiong LM, Schumaker KS, Zhu JK. Cell signaling during cold, drought, and salt stress. Plant Cell. 2002;**14**:S165-S183

[33] Seki M, Kameiy A, Yamaguchi-Shinozakiz K, Shinozaki K. Molecular responses to drought, salinity and frost: Common and different paths for plant protection. Current Opinion in Biotechnology. 2003;**14**:194-199

[34] Shanker AK, Maheswari M, Yadav SK, Desai S, Bhanu D, Attal NB, et al. Drought stress responses in crops. Functional and Integrative Genomics. 2014;**14**(1):11-22

[35] Golldack D, Li C, Mohan H, Probst N. Tolerance to drought and salt stress

in plants: Unraveling the signaling networks. Frontiers in Plant Science. 2014;**5**:151

[36] Wei K, Wang Y, Zhong X, Pan S. Protein kinase structure, expression and regulation in maize drought signaling. Molecular Breeding. 2014;**34**:583-602. DOI: 10.1007/s11032-014-0059-6

[37] Park SY, Fun P, Nishimura N, Jensen DR, Fujii H, Zhao Y, et al. Cuttler SRAbscisic acid inhibits type 2C protein phosphatases via the PYR/PYL family of START proteins. Science. 2009;**324**:1068-1071

[38] Geiger D, Maierhofer T, Al-Rasheid KA, Scherzer S, Mumm P, Liese A, et al. Stomatal closure by fast abscisic acid signaling is mediated by the guard cell anion channel SLAH3 and the receptor RCAR1. Science Signaling. 2011;**4**:ra32

[39] Raghavendra AS, Gonugunta VK, Christmann A, Grill E. ABA perception and signaling. Trends in Plant Science. 2010;**15**:595-401

[40] Chen WQJ, Zhu T. Networks of transcription factors with roles in environmental stress response. Trends in Plant Science. 2004;**9**:591-596. DOI: 10.1016/j.tplants.2004.10.007

[41] Gosal SS, Wani SH, Kang MS. Biotechnology and drought tolerance. Journal of Crop Improvement. 2009;**23**:19-54

[42] Gahlaut V, Jaiswal V, Kumar A, Gupta PK. Transcription factors involved in drought tolerance and their possible role in developing drought tolerant cultivars with emphasis on wheat (*Triticum aestivum* L.). Theoretical and Applied Genetics. 2016;**129**:2019-2042. DOI: 10.1007/s00122-016-2794-z

[43] Fujita Y, Nakashima K, Yoshida T, Katagiri T, Kidokoro S, et al. Three SnRK2 protein kinases are the main positive regulators of abscisic acid signaling in response to water stress in *Arabidopsis*. Plant & Cell Physiology. 2009;**50**:2123-2132

[44] Mizoi J, Ohori T, Moriwaki T, Kidokoro S, Todaka D, Maruyama K, et al. GmDREB2A;2, a canonical dehydration-responsive element-binding protein2-type transcription factor in soybean, is posttranslationally regulated and mediates dehydration-responsive element-dependent gene expression. Plant Physiology. 2013;**161**(1):346-361

[45] Li X, Zhang D, Li H, Wang Y, Zhang Y, Wood AJ. EsDREB, a novel truncated DREB2-type transcription factor in the desert legume *Eremosparton songoricum* enhances tolerance to multiple abiotic stresses in yeast and transgenic tobacco. BMC Plant Biology. 2014;**14**:44. DOI: 10.1186/1471-2229-14-44

[46] Li H, Zhang D, Li X, Guan K, Yang H. Novel DREB A-5 subgroup transcription factor from desert moss (*Syntrichia caninervis*) confers multiple abiotic stress tolerance to yeast. Journal of Plant Physiology. 2016;**194**:45-53

[47] Zou X, Seemann JR, Neuman D, Shen QJ. A WRKY gene from creosote bush encodes an activator of the abscisic acid signaling pathway. The Journal of Biological Chemistry. 2004;**279**(53):55770-55779

[48] Chen J, Xia X, Yin W. A poplar DRE-binding protein gene, PeDREB2L, is involved in regulation of defense response against abiotic stress. Gene. 2011;**483**:36-42

[49] Li H, Yao W, Fu Y, Li S, Guo Q. De novo assembly and discovery of genes that are involved in drought tolerance in Tibetan *Sophora moorcroftiana*. PLoS One. 2015;**10**(1):e111054. DOI: 10.1371/journal.pone.0111054

[50] McDowell NG. Mechanisms linking drought, hydraulics, carbon

metabolism, and vegetation mortality. Plant Physiology. 2011;**155**(3):1051-1059

[51] Chaves MM, Flexas J, Pinheiro C. Photosynthesis under drought and salt stress: Regulation mechanisms from whole plant to cell. Annals of Botany. 2009;**103**:551-560

[52] Van Breusegem F. L-10 plant oxidative stress signaling: Towards the proteome and beyond. Free Radical Biology and Medicine. 2018;**120**(supplement 1):S8

[53] Demidchick V. Mechanisms of oxidative stress in plants: From classical chemistry to cell biology. Environmental and Experimental Botany. 2015;**109**:212-228

[54] Bray EA. Classification of genes differentially expressed during water-deficit stress in *Arabidopsis thaliana*: An analysis using microarray and differential expression data. Annals of Botany. 2002;**89**:803-811

[55] Siddiqui NU, Chung H-J, Thomas TL, Drew MC. Abscisic acid-dependent and independent expression of the carrot late-embryogenesis-abundant-class gene Dc3 in transgenic tobacco seedlings. Plant Physiology. 1998;**118**:1181-1190

[56] Colmenero-Flores JM, Moreno LP, Smith CE, Covarrubias AA. Pvlea-18, a member of a new late-embryogenesis-abundant protein family that accumulates during water stress in the growing regions of well-irrigated bean seedlings. Plant Physiology. 1999;**120**:93-103

[57] Magwanga RO, Lu P, Kirungu JN, Lu H, Wang X, Cai X, et al. Characterization of the late embryogenesis abundant (LEA) proteins family and their role in drought stress tolerance in upland cotton. BMC Genetics. 2018;**19**:6. DOI: 10.1186/s12863-017-0596-1

[58] Gao J, Lan T. Functional characterization of the late embryogenesis abundant (LEA) protein gene family from *Pinus tabuliformis* (Pinaceae) in *Escherichia coli*. Scientific Reports. 2016;**6**:19467. DOI: 10.1038/srep19467

[59] Pedrosa AM, Martins CPS, Gonçalves LP, Costa MGC. Late embryogenesis abundant (LEA) constitutes a large and diverse family of proteins involved in development and abiotic stress responses in sweet orange (*Citrus sinensis* L Osb.). PLoS One. 2015;**10**(12):e0145785. DOI: 10.1371/journal.pone.0145785

[60] Hand SC, Menze MA, Toner M, Boswell L, Moore DLEA. Proteins during water stress: Not just for plants anymore. Annual Review of Physiology. 2011;**73**:115-134

[61] Dang NX, Popova AV, Hundertmark M, Hincha DK. Functional characterization of selected LEA proteins from *Arabidopsis thaliana* in yeast and in vitro. Planta. 2014;**240**(2):325-336

[62] Wang Y, Jiang J, Zhao X, Liu G, Yang C, Zhan L. A novel LEA gene from *Tamarix androsowii* confers drought tolerance in transgenic tobacco. Plant Science. 2006;**171**:655-662

[63] Gao W, Bai S, Li Q, Gao C, Liu G, Li G, et al. Overexpression of TaLEA gene from *Tamarix androsowii* improves salt and drought tolerance in transgenic poplar (*Populus simonii* x *P. nigra*). PLoS One. 2013;**8**(6):e67462. DOI: 10.1371/journal.pone.0067462

[64] Zhang J, Duan Z, Zhang D, Zhang J, Di H, Wu F, et al. Co-transforming bar and CsLEA enhanced tolerance to drought and salt stress in transgenic alfalfa (*Medicago sativa* L.). Biochemical and Biophysical Research Communications. 2016;**472**:75-82

[65] Kahn MS, Ahmad D, Kahn MA. Utilization of genes encoding

osmoprotectants in transgenic plants for enhanced abiotic stress tolerance. Electronic Journal of Biotechnology. 2015;**18**:257-266

[66] Gagneu D, Amouche A, Duhaze C, Lugan R, Larher FR, Bouchereau A. A reassessment of the function of the so-called compatible solutes in the halophytic *Plumbaginaceae* limonium latifolium. Plant Physiology. 2007;**144**:1598-1611

[67] Rabbani G, Choi I. Roles of osmolytes in protein folding and aggregation in cells and their biotechnological applications. International Journal of Biological Macromolecules. 2018;**109**:483-491

[68] Scholz FG, Bucci SJ, Arias N, Meinzer FC, Goldstein G. Osmotic and elastic adjustments in cold desert shrubs differing in rooting depth: Coping with drought and subzero temperatures. Oecologia. 2012;**170**:885-897. DOI: 10.1007/s00442-012-2368-y

[69] Ramadan AM, Hassanein SE. Characterization of P5CS gene in *Calotropis procera* plant from the de novo assembled transcriptome contigs of the high-throughput sequencing dataset. Comptes Rendus Biologies. 2014;**337**:683-690

[70] Vendruscoloa CG, Schusterb I, Pileggic M, Scapimd CA, Correa Molinarie HB, Marure CJ, et al. Stress-induced synthesis of proline confers tolerance to water deficit in transgenic wheat. Journal of Plant Physiology. 2007;**164**:1367-1376

[71] Quin D, Zhao C-L, Liu X-Y, Wang P-W. Transgenic soybeans expressing betaine aldehyde dehydrogenase from *Atriplex canescens* show increased drought tolerance. Plant Breeding. 2017;**136**:699-709. DOI: 10.1111/pbr.12518

[72] Wang X, Gao F, Bing J, Sun W, Feng X, Ma X, et al. Overexpression of the jojoba aquaporin gene, ScPIP1, enhances drought and salt tolerance in transgenic Arabidopsis. International Journal of Molecular Sciences. 2019;**20**:153. DOI: 10.3390/ijms20010153

[73] Noctor G, Lelarge-Trouverie C, Mhamdi A. The metabolomics of oxidative stress. Phytochemistry. 2015;**112**:33-53

[74] Farmer EE, Mueller MJ. ROS-mediated lipid peroxidation and RES-activated signaling. Annual Review of Plant Biology. 2013;**64**:429-450

[75] Das SK, Patra JK, Thatoi H. Antioxidative response to abiotic and biotic stresses in mangrove plants. A review. International Review of Hydrobiology. 2016;**101**:3-19. DOI: 10.1002/iroh.201401744

[76] Talbi S, Romero-Puertas MC, Hernández A, Terrón L, Ferchichi A, Sandalio LM. Drought tolerance in a saharian plant oudneya africana: Role of antioxidant defences. Environmental and Experimental Botany. 2015;**111**:114-126

[77] Wang L, Xu C, Wang C, Wang Y. Characterization of a eukaryotic translation initiation factor 5A homolog from *Tamarix androssowii* involved in plant abiotic stress tolerance. BMC Plant Biology. 2012;**12**:118. Available from: http://www.biomedcentral.com/1471-2229/12/118

[78] Du F, Shi H, Zhang X, Xu X. Responses of reactive oxygen scavenging enzymes, proline and malondialdehyde to water deficits among six secondary successional seral species in loess plateau. PLoS One. 2014;**9**(6):e98872. DOI: 10.1371/journal.pone.0098872

[79] Reddy PCO, Sairanganayukulu G, Thippeswamy M, Reddy PS, Reddy MK, Sudhakar C. Identification of stress-induced genes from the drought

tolerant semi-arid legume crop horse gram (*Macrotyloma uniflorum* (lam.) verdc) through analysis of subtracted expressed sequence tags. Plant Science. 2008;**175**:372-384

[80] Bhardwaj J, Gangwar I, Panzade G, Shankar R, Yadav SK. Global de novo protein-protein interactome elucidates interactions of drought proteins in horse gram (*Macrotyloma uniflorum*). Journal of Proteome Research. 2016;**15**:1794-1809. DOI: 10.1021/acs.jproteome.5b01114

[81] Zhou Y, Fei G, Ran L, Feng J, Li H. De novo sequencing and analysis of root transcriptome using 454 pyrosequencing to discover putative genes associated with drought tolerance in *Ammopiptanthus mongolicus*. BMC Genomics. 2012;**13**(7):1-17

[82] Wu Y, Wei W, Pang X, Wang X, Zhang H, Dong B, et al. Comparative transcriptome profiling of a desert evergreen shrub, *Ammopiptanthus mongolicus*, in response to drough and cold stresses. BMC Genomics. 2014;**15**:671

[83] Pang T, Guo L, Shim D, Cannon N, Tang S, Chen J, et al. Characterization of the transcriptome of the xerophyte *Ammopiptanthus mongolicus* leaves under drought stress by 454 pyrosequencing. PLoS One. 2015;**10**(8):e0136495. DOI: 10.1371/journal.pone.0136495

[84] Shi Y, Yan X, Zhao P, Yin H, Zhao X, Xiao H, et al. Transcriptomic analysis of a tertiary relict plant, extreme xerophyte Reaumuria soongorica to identify genes related to drought adaptation. PLoS One. 2013;**8**(5):e63993. DOI: 10.1371/journal.pone.0063993

[85] Liu Y, Liu M, Li X, Cao B, Ma X. Identification of differentially expressed genes in leaf of *Reaumuria soongorica* under PEG-induced drought stress by digital gene expression profiling. PLoS One. 2014;**9**(4):e94277. DOI: 10.1371/journal.pone.0094277

[86] Long Y, Zhang J, Tian X, Wu S, Zhang Q, Zhang J, et al. De novo assembly of the desert tree *Haloxylon ammodendron* (C.A. Mey.) based on RNA-Seq data provides insight into drought response, gene discovery and marker identification. BMC Genomics. 2014;**15**:1111. DOI: 10.1186/1471-2164-15-1111

[87] Zhang L, Hu X, Miao X, Chen X, Nan S, Fu H. Genome-scale transcriptome analysis of the desert shrub *Artemisia sphaerocephala*. PLoS One. 2016;**11**(4):e0154300. DOI: 10.1371/journal.pone.0154300

[88] Ma Q, Bao A-K, Chai W-W, Wang W-Y, Zhang J-L, Li Y-X, et al. Transcriptomic analysis of the succulent xerophyte *Zygophyllum xanthoxylum* in response to salt treatment and osmotic stress. Plant and Soil. 2016;**402**:343-361

[89] Muvunyi BP, Yan Q, Wu F, Min X, Yan ZZ, Kanzana G, et al. Mining late embryogenesis abundant (LEA) family genes in *Cleistogenes songorica*, a xerophyte perennial desert plant. International Journal of Molecular Sciences. 2018;**19**:3430. DOI: 10.3390/ijms19113430

[90] George S, Manoharan D, Li J, Britton M, Parida A. Transcriptomic responses to drought and salt in desert tree *Prosopis juliflora*. Plant Gene. 2017;**12**:114-122

Chapter 5

Canopy Temperature Depression as an Effective Physiological Trait for Drought Screening

Parvaze Ahmad Sofi, Asmat Ara, Musharib Gull and Khalid Rehman

Abstract

Water stress is a major production constraint in agriculture worldwide. Efforts to breed for drought tolerance are invariably hampered by the amount of time required to phenotype a large number of individuals and poor or inconsistent correlations and multiple mechanisms involved. Canopy temperature depression has emerged as a potential surrogate in view of substantial natural variation in crops as well as its correlation with yield. Based on the experimental findings two types of ideotype models based on CTD have been proposed as isohydric ("water saving") and anisohydric ("water spending"). The isohydrics have advantage in the harsher environments, whereas the anisohydrics perform better under moderate/mild drought situations. Water savers have a shallow root system with intermediate root growth and thin roots. They are early and have high water use efficiency, reduced transpiration and limited leaf area and canopy biomass development and superior photosynthate remobilization to pod and grain. Contrary to this, water spenders have a vigorous and deep rooting system with rapid root growth and a thicker root system. Such genotypes are early and have highly effective water use, moderate transpiration and fast leaf area and canopy biomass development, moderate sink strength and superior photosynthate remobilization to pod and grain formation.

Keywords: drought stress, physiology, canopy temperature depression

1. Introduction

The world is currently experiencing the combined effect of population growth and climate change leading to an unsustainable use of food and water resources. The population is going to touch 9.8 billion, and demand for food and feed crops is expected to almost double by 2050 [1]. Climate change models predict an increase in temperatures and increased frequency of severe events such as droughts and floods [2]. Higher temperatures may increase precipitation but also increase evaporation from crops, land, and surface water. An increase in the frequency of droughts and floods that invariably lead to crop failures can have a devastating effect on food availability and consequently accessibility. As per the current IPCC projections, the 20-year extreme annual daily maximum temperature will likely increase by about

1–3°C by the mid-twenty-first century and by about 2–5°C by the late twenty-first century, depending on the region and emission scenario [3]. Based on historical data collected in Africa on more than 20,000 trials between 1999 and 2007, each "degree day" spent above 30° is likely to reduce crop yields by 1% under optimal conditions and that penalty is going to increase up to 1.7% under water-limited conditions [4]. The impact of a climate change is not only about the projected increase in temperature, but it also affects the magnitude and distribution of rainfall, as well as availability of water at critical times of the crop growth [5]. While as the total amount of rain has recorded an increase in Africa over the last few years, the erratic and unpredictable nature of the drought and floods cycle has also increased [6].

Globally, rainfed agriculture is practiced in 80% of the total agricultural area and generates 62% of the world's staple food (FAOSTAT, 2011). In view of the current global water scarcity scenarios, climate change implications, and increases in demand for nonagricultural water use, the expansion of the area under irrigation, especially in developing countries, does not seem to be a realistic proposition to address food security challenges. Drought is one of the major production constraints in agriculture worldwide. It principally affects crops cultivated under rainfed conditions, which represent 80% of the total cultivated area worldwide. It is estimated that cultivation on the earth is only possible on 16% of the potentially arable area due to limited availability of water [7, 8]. Africa is strongly affected by drought almost every 12 years, but drought intensified during the years 2009–2011, during which, the wheat yields reduced by 45% in Kenya [9]. Similar trends have also been reported from Australia where drought reduced wheat yields by 46% in 2006 [10]. Around 17% of the global cultivated area was affected by drought during the period 1980–2006 [11]. **Tables 1** and **2** depict the proportion of cultivated areas implicated by drought stress and estimated yield reductions reported in various crops.

2. Breeding for drought tolerance

Breeding for drought tolerance is a sustainable option to reduce the risk of crop failure by improving the ability of crop plants to extract water from the deeper soil strata through better root architecture, by decreasing the amount of crop water demands (improving water use efficiency), or by improving a crop's ability to survive longer periods without water, thereby ultimately increasing yields in rainfed environments. However, breeding for drought tolerance is complex because

Region	Crop species	Proportion of the cultivated area affected by drought
Africa	Wheat	80%
Eastern Asia	Maize	50%
Europe	Maize	60%
North America	Wheat	47%
Oceania	Barley	70%
South America	Maize	50%
South Asia	Wheat	65%
Southeast Asia	Rice	65%

Table 1.
Proportion of the cultivated area affected by drought in different regions and for different crops (source: [12]).

Crop	Yield reduction	References
Barley	49–57%	[13]
Chickpea	78%	[14]
Groundnut	55–72%	[15]
Maize	43–80%	[16]
Oat	79%	[17]
Potato	89%	[17]
Rice	42–66%	[18]
Brassica	39%	[17]
Rye	52%	[17]
Bread wheat	57%	[17]
Durum wheat	74%	[19]
Pigeon pea	42%	[20]
Green gram	71%	[20]
Black gram	74%	[20]

Table 2.
Impact of drought stress on yield reduction in different crops.

it involves quantitative inheritance and environmental influence [21]. Efforts to breed for drought tolerance are invariably hampered by the amount of time required to phenotype a large number of individuals and poor or inconsistent correlation between a phenotype and yield under drought conditions due, in part, to multiple mechanisms involved. Various authors have investigated the genetic basis of drought tolerance in common bean and reported that both additive and nonadditive gene actions are involved in drought tolerance [22, 23]. Schneider et al. [24] reported a strong genotype x environment interaction in the expression of identified quantitative trait locus (QTL), such that potential for marker-assisted selection in breeding for drought tolerance was also inconclusive. Selection based solely on yield performance confounds the complexity of breeding for drought as yield is a highly complex trait with low heritability especially under stress conditions. Therefore, it is imperative to identify less complex traits related to the drought that will improve upon selection for drought tolerance and separate these traits into major components which may help further understanding of the genetic basis.

A better understanding of the relationship between below- and aboveground traits will contribute to improved productivity under drought stress. Root traits including structure and their spatial distribution of root system in different soil horizons are essential for yield improvement because of its high heritability under drought stress [25–27] and high correlation with yield traits [28]. However, extensive use of roots as the target traits for developing climate resilience suffers from the difficulties associated with studying roots, especially under field conditions. The shoot traits are easy to measure and quantify; however, it has to be linked with root traits with the perspective of improving drought tolerance. In the following sections, we discuss some of the potential aboveground traits that have been shown to be correlated with improved drought tolerance as well as better grain yield under stress. Currently, there is a huge shopping list of relatively unranked traits that have been proposed to be used as surrogates for drought tolerance response. Canopy temperature depression has emerged as a potential surrogate in view of substantial natural variation in crops as well as its correlation with yield under both stress and nonstress conditions [29].

3. Canopy temperature (CT) and canopy temperature depression (CTD)

Plant water balance is a direct measure of drought response of crops. In fact, the transpiration is the main cause of changes in leaf temperature, and there is a direct relationship between leaf temperature, transpiration rate, leaf porosity, and stomatal conductance [30]. As long as the plants continue to transpire through open stomata, the canopy temperatures could be maintained at metabolically comfortable range; otherwise, higher temperature would slow or retard the vital enzymatic activities and consequently the overall metabolism. The closure of stomata for a considerable period of time, especially during the periods of higher evaporative demands driven by high temperature and vapor pressure deficit, is known to increase the leaf temperature [31] and hamper plant's ability to maintain a relatively cooler canopy during grain filling period as an important physiological adaptation for stress [22]. Canopy temperature differences have been shown to correlate well with the transpiration status in rice, potatoes, wheat, and sugar beet. Deviation of temperature of plant canopies from the ambient temperature, also known as canopy temperature depression = air temperature (Ta) − canopy temperature, has been recognized as an indicator of overall plant water status [33] and facilitates in evaluation of plant response to stresses like high temperature [34] and drought [35, 36]. CTD is positive when the canopy is cooler than the air, and this value has been associated with yield increase in different crops [37, 38]. The thermal imagery system is a powerful tool as it can capture the temperature difference of plant canopies quite rapidly.

Thermal infrared imaging and infrared thermography (IRT), to measure the canopy or leaf temperature, are the twin approaches that measure the extent of evaporative cooling occurring in a crop canopy and allow a remote sensing of the plant water balance. Between these two approaches, thermal infrared imaging through an infrared camera offers several benefits compared with temperature sensors, most importantly the facility for spatial resolution and the ability to sample larger area. Most infrared cameras currently have arrays of 320 × 240 sensor elements, which mean that >75,000 individual temperature readings are recorded in a single image. This allows more precise measurements in a fraction of the time needed to perform several replicate readings per plot, which is also prone to error due to changing environmental conditions between measurements. Canopy temperature is one such integrative trait that reflects the plant water status or the resultant equilibrium between root water uptake and shoot transpiration [39]. Canopy temperature has been used successfully as selection criteria in breeding for drought-prone environments [33, 37, 40].

At ambient temperature, all objects emit far-infrared light of approximately 10 μm wavelength [41]. Detectors sensitive in the 8–14 μm wavelength bands convert this radiation into a temperature reading. Such detectors are the basis of non-imaging infrared thermometers, which yield an average temperature measurement of all objects within the field of view. Applications of these simple and affordable instruments include forest canopy studies and irrigation scheduling in field crops [42]. There are yet other thermometers based on infrared imaging that can capture images by adding a scanning system, and each point of measurement is a temperature value based on a pseudo-color value that depends on the radiation captured. The radiation is converted to visual pseudo-color images representing different temperature levels. Both the non-imaging and scanning image thermometers are now being routinely used to measure the temperatures of leaves or canopy in controlled and field conditions. In case of greenhouse or growth chamber experiments where only one or two plants are used per replication, leaf temperatures are used, whereas in case of field experiments where comparatively larger plots are used, canopy temperature is mostly used. Nowadays, unmanned aerial vehicles (UAVs) or robotic equipments fitted with sensors and cameras can be used for monitoring stress advancement in greenhouses and field trials.

Infrared thermometry was first used for scheduling crop irrigation in the 1970s [43], while the use of canopy temperature in drought screening began in the early 1980s [44]. The use of canopy temperature in Centro Internacional de Mejoramiento de Maíz y Trigo or International Maize and Wheat Improvement Center (CIMMYT) breeding research began in the early 1990s for hot, irrigated environments [45] and has also been used as a selection criterion for isolating drought-tolerant parental lines for initiating strategic crossing as well as for early generation selection under drought (i.e., from F_3 generation onward). Canopy temperature measured by non-imaging IR thermometer can markedly accelerate selection of drought-tolerant genotypes given on high operational speed (≈10 seconds per plot), simplicity, and relatively economically friendly measurements. It is also integrative of the whole canopy due to scoring many plants at once, thus reducing error associated with plant-to-plant variation [46]. In addition, measurements of CT on plants do not interfere with the sensitive stomata, in comparison with other methods that estimate leaf conductance such as porometry and other gas exchange approaches. These may include accurate estimation of the temperatures of different organs of a single plant or the simultaneous capture of CT of all plots in a large trial [47, 48]. Besides, canopy temperature may be related directly to the genetic potential of the root's capacity to explore soil moisture [49]; however, factors such as microelement deficiency or soil-borne disease that affect root growth may confound the relationship.

Grant et al. [50] investigated the robustness and sensitivity of thermal imaging for detecting changes in stomatal conductance and leaf water status in a range of plant species (grapevine, bean, and lupin) under greenhouse or controlled environment conditions. In particular, they compared absolute leaf temperatures and thermal indices of plant stress with stomatal conductance and water potential. Thermal imaging is successfully distinguished between irrigated and nonirrigated plants of different species, with strong correlations between thermal indices and stomatal conductance as measured with a leaf pyrometer. Factors such as leaf angle are important and should be given due consideration when using thermal imaging for indirect measurement of the level of drought stress of the tested materials (**Figure 1**).

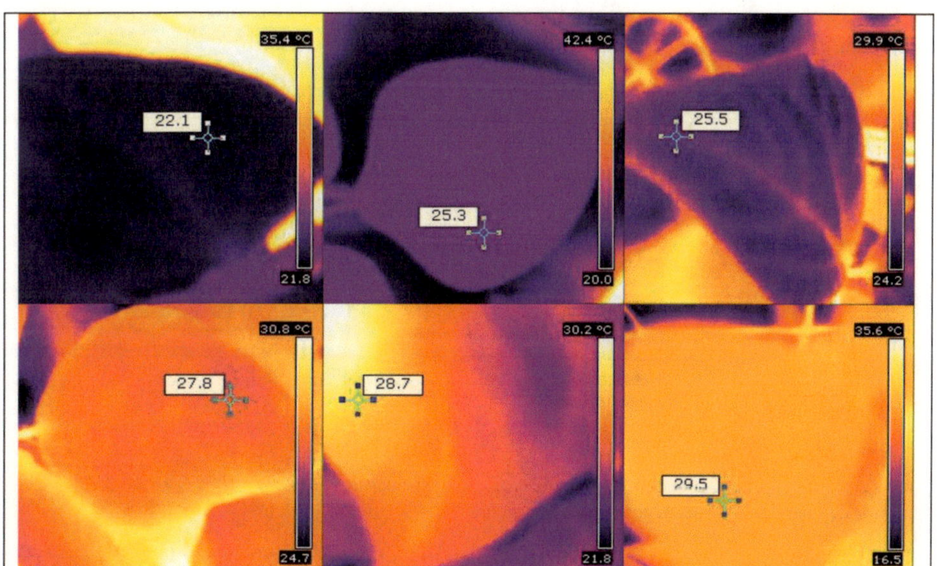

Figure 1.
Infrared camera images of bean leaves (source: P. A. Sofi).

4. CTD as an effective surrogate trait for drought screening

Canopy temperature is one of the many physiological traits that may help to identify drought-tolerant cultivars. Canopy temperature depression is the difference between air temperature and plant canopy temperature [51]. Under drought conditions, stomatal conductance decreases when soil moisture is not adequate to keep up with evaporative demands; and this, in turn, increases canopy temperature [52]. Plant morphological trait such as canopy architecture also influences canopy temperature not only through the angle of leaves to the light source but also through the degree of mutual shading in the canopy . Canopy temperature can provide plant-based information on the water status of the crop [53]. Under both greenhouse and field conditions, genotypes with a cooler canopy temperature (higher CTD) under drought stress use more available soil moisture to cool the canopy by transpiration to avoid excessive dehydration [54, 55]. In a large number of experiments in diverse crops, CTD has been found to have significant correlation with grain yield (**Table 3**).

Canopy temperature is also related directly to the genetic potential of the root's capacity to explore soil moisture [32, 56]. Canopy temperature depression can be used as effective proxy traits for the analysis of root development and biomass partitioning under drought stress [57]. Cool canopy temperatures are reported to be associated with enhanced plant access to water by virtue of deeper roots [49], and the common bean genotypes with cooler canopy temperatures reported 30% more yield associated with an increase of 40% in root dry weight at 60–120 cm. Canopy temperature depression has been shown to be correlated with yield under drought stress ([32, 35, 58, 59]; **Table 3**) and hot irrigated conditions [32, 60]. Canopy temperatures under well-watered conditions also indicate potential yield performance during drought and could effectively be used as a technique to assess genotypic response to drought [61]. Blum et al. [62] used canopy temperatures of drought stress wheat genotypes to characterize yield stability under various moisture conditions. A positive correlation was found between a drought susceptibility index and canopy temperature in stressed environments. Drought-susceptible genotypes which suffered relatively greater yield loss under drought stress tended to have warmer canopies at midday.

Crop	Trait relationship with yield	References
Wheat	Positive association (r^2 = 0.45–0.89; $P \leq 0.001$)	Amani et al. [60]; Fischer et al. [37]; Balota et al. [33]
Chickpea	Positive association (r^2 = 0.40; $P \leq 0.001$)	Purushothaman et al. [59]
Common bean	Positive association (r^2 = 0.11–0.32; $P \leq 0.001$)	Asfaw et al. [25] and Polania et al. [28]
Groundnut	Positive association (r^2 = 0.44; $P \leq 0.001$)	Singh et al. [67]
Sorghum	Positive association (r^2 = 0.19; $P \leq 0.001$)	Mutawa [58]
Triticale and barley	Positive association (r^2 = 0.76; $P \leq 0.001$)	Roohi et al. [68]

Table 3.
Correlation of CTD with grain yield in various crops.

CTD can be affected by biological and environmental factors like water status of soil, wind, evapotranspiration, cloudiness, conduction systems, plant metabolism, air temperature, relative humidity, and continuous radiation [63] and has preferably been measured in high air temperature and low relative humidity because of high vapor pressure deficit conditions [60]. At the end of the 1980s, CIMMYT began CTD measurements on different irrigated experiments in Northwest Mexico. Phenotypic correlations of CTD with grain yield were occasionally positive [37]. CTD has been used as selection criteria for tolerance to drought and high temperature stress in wheat breeding, and the used breeding method is generally mass selection in early generations like F3. According to this method, firstly, bulks which show high CTD value (have cool canopy) are selected in F3 generation. Later, single plants which show high stomata conductance (g) among bulks also show cool canopy at the same selection generation; thus, both of these traits are used at the same breeding program [63].

CTD can be a reliable indicator of crop performance under both irrigated and drought stress conditions. Under irrigated conditions there was a linear trend of higher yield with CTD; however, under drought stress, both negative CTD and positive CTD could be identified, and in both classes, high-yielding genotypes were identified. The water savers probably could sense drought stress in early phases of growth and could trigger conservative water use that could be used in later stages of growth [30]. However, the reduction in water use is generally achieved by plant traits and environmental responses that could also reduce yield potential [64]. Under optimum experimental conditions provided that data are collected when the canopy is sufficiently expanded to cover the soil, CTD can be a good predictor of crop yield (r = 0.6–0.85; [65]). In wheat, yield progress was found to be associated with cooler canopies [37], and significant genetic gains in yield have been reported in response to direct selection for CTD [55, 65]. Reynolds et al. have made a comparative analysis of aerial and handheld IR thermometers and found that correlation of CTD with grain yield was comparable (r = 0.68** and 0.73**, respectively).

5. Toward a crop ideotype based on canopy temperature depression

Blum has proposed ideotypes of crop plants based on canopy temperature depression for use in plant breeding as per the drought types such as the isohydric ("water saving") model and the anisohydric ("water spending") model. The water saving model has a distinct advantage in the harsher environments, whereas the water spending model is expected to perform relatively better under more moderate/mild drought situations. Polania et al. [28] have proposed that the water spender genotypes can be used for cultivation in areas exposed to intermittent drought stress with soils that can store greater amount of available water deep in the soil profile. However, water savers can be more suitable in semiarid to dry environments dominated by the terminal drought stress. The water savers or isohydric genotypes are characterized by a shallow root system with intermediate root growth and penetration ability and thin roots. Such genotypes are early and have high water use efficiency, reduced transpiration and limited leaf area and canopy biomass development, reduced sink strength, and superior photosynthate remobilization to pod and grain formation. Contrary to this, water spenders or anisohydric genotypes have a vigorous and deep rooting system with rapid root growth rate and penetration ability and a thicker root system. Such genotypes are early and have highly effective water use, moderate transpiration and fast leaf area and canopy biomass development, moderate sink strength, and superior photosynthate remobilization to pod and grain formation.

Our studies in beans and cowpea have also revealed that CTD measurements can be used to build a crop ideotype for water stress response. In our studies with beans and cowpea, we found that CTD values across stages decreased progressively on account of rapid depletion of moisture (**Figure 2**). The genotypes could be grouped into water savers and water spenders using the sign of CTD values (**Figures 3** and **4**). The water spenders have higher stomatal conductance and lose water through transpiration, whereas water savers have conservative water use on account of lower stomatal conductance or early closure of stomata and as such have hotter canopies. Under irrigated conditions also, we found a linear relationship with genotypes having higher CTD values showing better yields, whereas under water-stressed conditions, high-yielding genotypes could be found in both groups.

Canopy temperature can be related to the genetic potential of the root's capacity to explore soil moisture [32, 56] and as such can be used as effective surrogate trait for the analysis of root development and biomass partitioning under drought stress [57]. Cool canopies (+CTD) are reported to be associated with enhanced plant access to water by virtue of deeper roots (Lopes and Reynolds 2010), and the genotypes with cooler canopies have been reported to yield 30% more, with a concomitant increase of 40% in root dry weight. CTD has been reported to be correlated with yield under both drought stress [32, 35, 59] and hot irrigated conditions [32]. Drought-susceptible genotypes which suffered relatively greater yield loss under drought stress tended to have warmer canopies at midday. Our studies have revealed that CTD can be a reliable indicator of crop performance under both irrigated and drought stress conditions. Under irrigated conditions, there was a linear trend of higher yield with CTD; however, under drought stress, both negative CTD and positive CTD could be identified, and in both classes, high-yielding genotypes were identified. The water savers probably could sense drought stress in early phases of growth and could trigger conservative water use that could be used in later stages of growth [30]. However, the reduction in water use is generally achieved by plant traits and environmental responses that could also reduce yield potential [64].

In recent years, with the availability of high-throughput phenotyping platforms, canopy temperature depression has been widely used to study genotypic response to drought. Blum et al. [62] used canopy temperatures of drought stress wheat genotypes to characterize yield stability under various moisture conditions. In most of the studies using CTD, a positive correlation has been found between

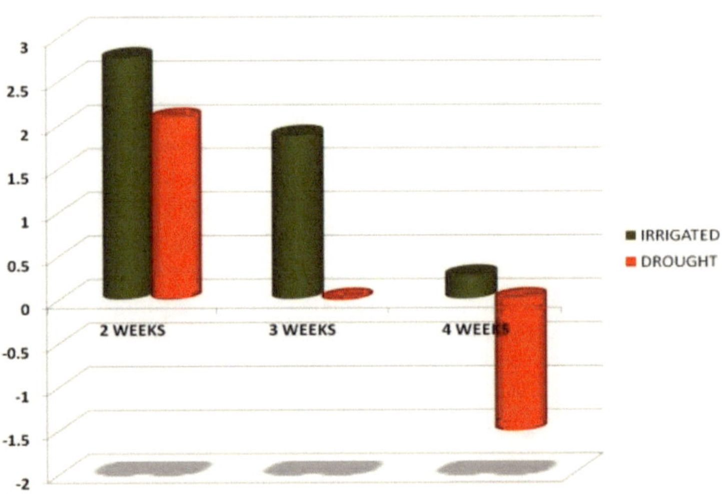

Figure 2.
Mean CTD across genotypes at the second, third, and fourth week of stress imposition.

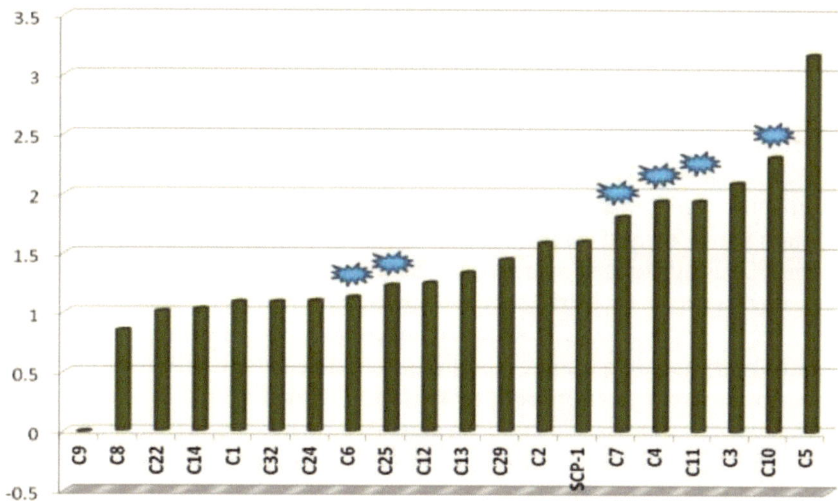

Figure 3.
Variation for CTD averaged over 3 stages in 20 genotypes of cowpea under irrigated conditions.

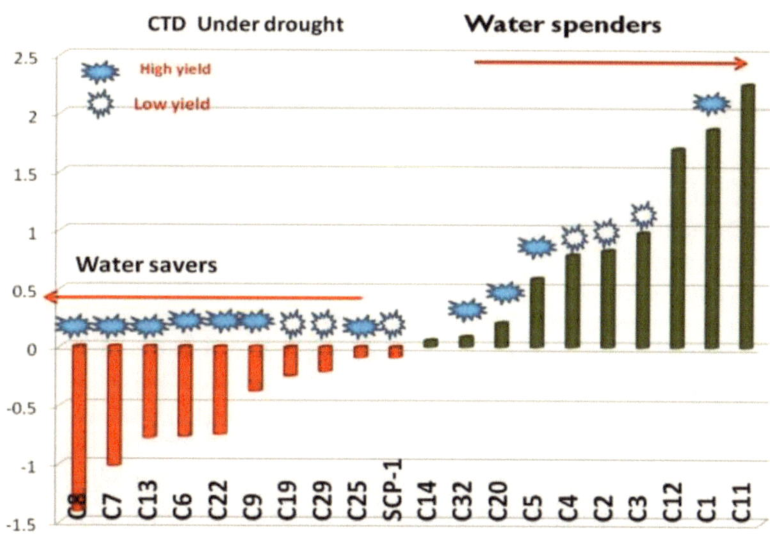

Figure 4.
Variation for CTD averaged over 3 stages in 20 genotypes of cowpea under drought stress.

drought susceptibility index and canopy temperature in stressed environments. Drought-susceptible genotypes which suffered relatively greater yield loss under stress tended to have warmer canopies at midday. Under well-watered conditions also, CTD provides a fair indication of potential yield performance during drought and could effectively be used as a technique to assess genotypic response to drought. Rashid et al. [35] reported that significant correlation between canopy temperature and yield under moisture-stress conditions and stress susceptibility index values indicated the potential for screening wheat genotypes for drought response. Canopy temperature depression is positive when the canopy is cooler than the air (CTD = Ta – Tc). It has been used in various practical applications including evaluation of plant response to environmental stress [66] and irrigation scheduling [69], to evaluate cultivars for water use [70], tolerance to heat [71], and drought [35, 62]. In general, CTD has been used to assess plant water status because it represents an overall, integrated physiological response to drought and high temperature [60]. Overall, the existing literature suggests that dominant mechanisms that increase CTD vary with environment and crop species .

Canopy temperature is a useful indicator of crop water status [43] and has the potential as a tool for indirect selection of genotypes tolerant to drought and heat-stressed environments [55]. For field experiments in wheat, CT data is most commonly measured on a whole-plot basis using a handheld infrared thermometer [71], although more rapid assessment using thermal imaging [72] is growing in popularity. CT is influenced by a number of environmental factors including the amount of solar radiation hitting the canopy, soil moisture, wind speed, temperature, and relative humidity [73]. Genetic differences in CT result from variation in the plant's ability to move water through the vascular system, differences in stomata aperture driving transpiration, root biomass and depth, metabolism, and source sink balance [74]. As such, CT has been shown to correlate with these physiological traits under field conditions and integrates them into a single low-cost diagnostic measurement that has a potential for selection of tolerant parental genotypes or early generation breeding lines [55]. CT has moderate heritability across environments in both diverse sets of germplasm [49] and in related material such as recombinant inbred populations [73]. Lopes and Reynolds [49] found similar broad-sense heritability for a diverse set of 294 spring wheat lines (H2 = 0.38) and a set of 169 sister lines (H2 = 0.34) across well-watered, drought-stressed, and heat-stressed environments in Northwest Mexico. Genetically, CT is a quantitative trait. Pierre et al. [74] determined the gene action for CT to be mainly additive by additive in five wheat populations with some dominant effects. Genetic mapping shows CT to be controlled mostly by small effect loci that are pleiotropic with variation in other traits, such as days to heading and plant height [20]. The correlation between CT and yield is consistently negative in the literature in both drought and heat environments such that a cooler canopy provides a yield benefit under stress [73]. Exceptions have been shown in both bread wheat [75], where CT measurements taken in Mexico were positively correlated with yield at international sites, and in durum wheat [76], where CT was found to increase with date of cultivar release and increasing yield. Experiments investigating CT are often conducted with sets of lines preselected for variation in canopy temperature or other tolerance traits [49], international trials of elite drought and heat tolerant lines [45], or using historical germplasm [9, 19, 21] and may not be representative of variation present in the early stages of yield testing in a breeding program. Reynolds et al. [55] demonstrated that advanced lines derived from "physiological crosses" targeted at one or more adaptive traits had a definite yield advantage over "conventional crosses" where physiological traits including CTD were not considered in parental selection. However, there is a need to investigate the ability of CT to select high-yielding lines within the germplasm flow of a breeding program where very little preselection for stress tolerance per se has been done.

6. Conclusion

Both empirical breeding and analytical approaches are used for improving crop performance under changing climate (drought, high temperature, etc.). However, there is a strong argument evolving in support of the analytical approaches based on indirect selection approaches using efficient surrogate traits to enhance the scale and reliability of phenotyping. Infrared thermometry can detect small differences in leaf temperature in both field and greenhouse conditions, measurements are fast and nondestructive, and the trait has a moderate to high heritability and shows positive correlation with yield [44]. Measurements should however be made well before the crop maturity and due consideration should be given to biological and environmental factors such as water status of soil, wind, evapotranspiration, cloudiness, conduction systems, plant metabolism, air temperature, relative humidity, and continuous radiation [55]. In light of substantial experimental evidence that a fairly positive relationship exists between yield and CTD under both stressed and nonstressed conditions, it is essential to incorporate CTD as effective complementary trait in selection programs aimed at developing climate resilient varieties.

Acknowledgements

The facilities provided by the Faculty of Agriculture are gratefully acknowledged.

Author details

Parvaze Ahmad Sofi*, Asmat Ara, Musharib Gull and Khalid Rehman
Dry Land Agricultural Research Station, SKUAST-Kashmir, India

*Address all correspondence to: parvazesofi@gmail.com

IntechOpen

© 2019 The Author(s). Licensee IntechOpen. This chapter is distributed under the terms of the Creative Commons Attribution License (http://creativecommons.org/licenses/by/3.0), which permits unrestricted use, distribution, and reproduction in any medium, provided the original work is properly cited.

References

[1] Foley JA, Ramankutty N, Brauman KA, Cassidy ES, Gerber JS, Johnston M, et al. Solutions for a cultivated planet. Nature. 2011;**478**:337-342. DOI: 10.1038/nature10452

[2] Kundzewicz ZW, Mata LJ, Arnell NW, Döll P, Jimenez B, Miller K, et al. The implications of projected climate change for freshwater resources and their management. Hydrological Sciences Journal. 2008;**53**(1):3-10

[3] IPCC. Chapter 3. In: Field CB, Barros V, Stocker TF, Qin D, Dokken DJ, Ebi KL, et al., editors. Changes in Climate Extremes and their Impacts on the Natural Physical Environment. Cambridge: Cambridge University Press; 2012. pp. 109-230

[4] Lobell DB, Bänziger M, Magorokosho C, Vivek B. Nonlinear heat effects on African maize as evidenced by historical yield trials. Nature Climate Change. 2011;**1**:42-45. DOI: 10.1038/nclimate1043

[5] Feng X, Porporato A, Rodriguez-Iturbe I. Changes in rainfall seasonality in the tropics. Nature Climate Change. 2013;**3**:811-815. DOI: 10.1038/NCLIMATE1907

[6] Douglas I, Alam K, Maghenda M, McDonnell Y, McLean L, Campbell J. Unjust waters: Climate change, flooding and the urban poor in Africa. Environment and Urbanization. 2008;**20**:187-205. DOI: 10.1177/0956247808089156

[7] Alexandratos N, Bruinsma J. World Agriculture Towards 2030/2050: The 2012 Revision (Vol. 12, No. 3). FAO, Rome: ESA Working paper; 2012

[8] Ondrasek G. Water scarcity and water stress in agriculture. In: Ahmad P, Wani M, editors. Physiological Mechanisms and Adaptation Strategies in Plants under Changing Environment. New York, NY: Springer; 2014

[9] Rojas R, Feyen L, Dosio A, Bavera D. Improving pan-European hydrological simulation of extreme events through statistical bias correction of RCM-driven climate simulations. Hydrology and Earth System Sciences Discussions. 2011;**8**:3883-3936

[10] van Dijk AI, Beck HE, Crosbie RS, de Jeu RA, Liu YY, Podger GM, et al. The millennium drought in Southeast Australia (2001-2009): Natural and human causes and implications for water resources, ecosystems, economy, and society. Water Resources Research. 2013;**49**(2):1040-1057

[11] Dai A. Increasing drought under global warming in observations and models. Nature Climate Change. 2013;**3**:52-58

[12] Li Y, Wei Y, Meng W, Xiadong Y. Climate change and drought: A risk assessment of crop yield impacts. Climate Research. 2009;**39**:31-46

[13] Samarah NH. Effects of drought stress on growth and yield of barley. Agronomy for Sustainable Development. 2005;**25**(1):145-149

[14] Szilagyi L. Influence of drought on seed yield components in common bean. Bulgarian Journal of Plant Physiology. 2003;**2003**:320-330

[15] Hamidou F, Halilou O, Vadez V. Assessment of groundnut under combined heat and drought stress. Journal of Agronomy and Crop Science. 2013;**199**(1):1-11

[16] Khalili M, Naghavi MR, Aboughadareh AP, Rad HN. Effects of drought stress on yield and yield

components in maize cultivars (*Zea mays* L.). International Journal of Agronomy and Plant Production. 2013;**4**(4):809-812

[17] Hlavinka P, Trnka M, Semerádová D, et al. Effect of drought on yield variability of key crops in Czech Republic. Agricultural and Forest Meteorology. 2009;**149**(3):431-442

[18] Raman A, Verulkar SB, Mandal NP, et al. Drought yield index to select high yielding rice lines under different drought stress severities. Rice. 2012;**5**(1):31

[19] Kiliç H, Yağbasanlar T. The effect of drought stress on grain yield, yield components and some quality traits of durum wheat (Triticum turgidum ssp. durum) cultivars. Notulae Botanicae Horti Agrobotanici Cluj-Napoca. 2010;**38**(1):164-170

[20] Kulkarni A, Gadgil S, Patwardhan S. Monsoon variability, the 2015 Marathwada drought and rainfed agriculture. Current Science. 2016;**111**:1182-1193

[21] Kashiwagi J, Krishnamurthy L, Upadhyaya HD, Krishna H, Chandra S, Vadez V, et al. Genetic variability of drought-avoidance root traits in the mini-core germplasm collection of chickpea (*Cicer arietinum* L.). Euphytica. 2005;**146**:213-222

[22] Munjal R, Rana RK. Evaluation of physiological traits in wheat (*Triticum aestivum* L.) for terminal high temperature tolerance. In: Proceedings of the Tenth International Wheat Genetics Symposium. Vol. 2. Paestum, Italy; 2003. pp. 804-805

[23] Fukuoka A, Kimura JI, Oshio T, Sakamoto Y, Ichikawa M. Preferential oxidation of carbon monoxide catalyzed by platinum nanoparticles in mesoporous silica. Journal of the American Chemical Society. 2007;**129**(33):10120-10125

[24] Schneider KA, Rosales-Serna R, Ibarra-Perez F, Cazares-Enriquez B, Acosta-Gallegos JA, Ramirez-Vallejo P, et al. Improving common bean performance under drought stress. Crop Science. 1997;**37**(1):43-50

[25] Asfaw A, Blair MW, Struik PC. Multienvironment quantitative trait loci analysis for photosynthate acquisition, accumulation, and remobilization traits in common bean under drought stress. G3: Genes, Genomes, Genetics. 2012;**2**(5):579-595

[26] Riaz ATIF, Younis A, Taj AR, Karim A, Tariq U, Munir S, et al. Effect of drought stress on growth and flowering of marigold (*Tagetes erecta* L.). Pakistan Journal of Botany. 2013;**45**(S1):123-131

[27] Khalid, R. Root Architecture and Biomass Partitioning in Relation to Drought Stress in Common Bean. MSc Thesis Submitted to SKUAST-K; 2017. p. 173

[28] Polania JA, Poschenrieder C, Beebe S, Rao IM. Effective use of water and increased dry matter partitioned to grain contribute to yield of common bean improved for drought resistance. Frontiers in Plant Science. 2016;**7**:660

[29] Pinto RS, Reynolds MP. Common genetic basis for canopy temperature depression under heat and drought stress associated with optimized root distribution in bread wheat. Theoretical and Applied Genetics. 2015;**128**(4):575-585

[30] Rebetzke GJ, Rattey AR, Farquhar GD, Richards RA, Condon AG. Genomic regions for canopy temperature and their genetic association with stomatal conductance and grain yield in wheat. Functional Plant Biology. 2013;**40**:14-33

[31] Mir RR, Zaman-Allah M, Sreenivasulu N, Trethowan R, Varshney RK. Integrated genomics, physiology and breeding approaches for improving drought tolerance in crops. Theoretical and Applied Genetics. 2012;**125**(4):625-645

[32] Blair MW, González LF, Kimani PM, Butare L. Genetic diversity, inter-gene pool introgression and nutritional quality of common beans (Phaseolus vulgaris L.) from Central Africa. Theoretical and Applied Genetics. 2010;**121**(2):237-248

[33] Balota M, Payne WA, Evett SR, Peters TR. Morphological and physiological traits associated with canopy temperature depression in three closely related wheat lines. Crop Science. 2008;**48**(5):1897-1910

[34] Reynolds MP, Skovmand B, Trethowan R, Pfeiffer W. Evaluating a conceptual model for drought tolerance. In: Ribaut JM, Poland D, editors. Molecular Approaches for the Improvement of Cereals for Stable Production in Water-Limited Environments. A Strategic Planning Workshop. 2000. pp. 49-53

[35] Rashid A, Stark JC, Tanveer A, Mustafa T. Use of canopy temperature measurements as a screening tool for drought tolerance in spring wheat. Journal of Agronomy and Crop Science. 1999;**182**(4):231-238

[36] Royo C, Villegas D, Del Moral LG, Elhani S, Aparicio N, Rharrabti Y, et al. Comparative performance of carbon isotope discrimination and canopy temperature depression as predictors of genotype differences in durum wheat yield in Spain. Australian Journal of Agricultural Research. 2002;**53**(5):561-569

[37] Fischer RA, Rees D, Sayre KD, Lu ZM, Condon AG, Saavedra AL. Wheat yield progress associated with higher stomatal conductance and photosynthetic rate, and cooler canopies. Crop Science. 1998;**38**:1467-1475

[38] Singh P, Kanemasu ET. Leaf and canopy temperatures of pearl millet genotypes under irrigated and nonirrigated conditions. Agronomy Journal. 1983;**75**:497-501

[39] Berger B, Parent B, Tester M. High-throughput shoot imaging to study drought responses. Journal of Experimental Botany. 2010;**61**(13):3519-3528

[40] Jones HG, Serraj R, Loveys BR, Xiong L, Wheaton A, Price AH. Thermal infrared imaging of crop canopies for the remote diagnosis and quantification of plant responses to water stress in the field. Functional Plant Biology. 2009;**36**(11):978-989

[41] North GB, Nobel PS. Changes in hydraulic conductivity and anatomy caused by drying and rewetting roots of Agave deserti (Agavaceae). American Journal of Botany. 1991;**78**:906-915

[42] Ranjbarfordoei A, Samson R, Van Damme P, Lemeur R. Effects of drought stress induced by polyethylene glycol on pigment content and photosynthetic gas exchange of *Pistacia khinjuk* and *P. Mutica*. Photosynthetica. 2000;**38**(3):443-447

[43] Jackson RD, Reginato RJ, Idso SB. Wheat canopy temperature—Practical tool for evaluating water requirements. Water Resources Research. 1977;**13**:651-656

[44] Blum A, Mayer J, Gozlan G. Infrared thermal sensing of plant canopies as a screening technique for dehydration avoidance in wheat. Field Crops Research. 1982;**5**:137-146

[45] Reynolds MP, Balota M, Delgado MIB, Amani I, Fischer RA. Physiological

and morphological traits associated with spring wheat yield under hot, irrigated conditions. Australian Journal of Plant Physiology. 1994;**21**:717-730

[46] Cossani CM, Reynolds MP. Physiological traits for improving heat tolerance in wheat. Plant Physiology. 2012;**160**(4):1710-1718

[47] Jones HG, Serraj R, Loveys BR, Xiong L, Wheaton A, Price AH. Thermal infrared imaging of crop canopies for the remote diagnosis and quantification of plant responses to water stress in the field. Functional Plant Biology. 2009;**36**(11):978-989

[48] Munns R, James RA, Sirault XR, Furbank RT, Jones HG. New phenotyping methods for screening wheat and barley for beneficial responses to water deficit. Journal of Experimental Botany. 2010;**61**(13):3499-3507

[49] Lopes MS, Reynolds MP. Partitioning of assimilates to deeper roots is associated with cooler canopies and increased yield under drought in wheat. Functional Plant Biology. 2010;**37**(2):147-156

[50] Grant OM, Chaves MM, Jones HG. Optimizing thermal imaging as a technique for detecting stomatal closure induced by drought stress under greenhouse conditions. Physiologia Plantarum. 2006;**127**(3):507-518

[51] Tuberosa R. Phenotyping for drought tolerance of crops in the genomics era. Frontiers in Physiology. 2014;**3**:347

[52] Urban O, Janous D, Acosta M, Czerny R, Marková I, Navratil M, et al. Ecophysiological controls over the net ecosystem exchange of mountain spruce stand. Comparison of the response in direct vs. diffuse solar radiation. Global Change Biology. 2007;**13**:157-168

[53] Mahan F, Isazadeh A, Khanli LM. Using an active fuzzy ECA rule-based negotiation agent in e-commerce. International Journal of Electronic Commerce Studies. 2011;**2**(2):127-148

[54] Ludlow MM, Muchow RC. A critical evaluation of traits for improving crop yields in water limited environments. Advances in Agronomy. 1990;**43**:107-153

[55] Reynolds M, Manes Y, Izanloo A, Langridge P. Phenotyping approaches for physiological breeding and gene discovery in wheat. The Annals of Applied Biology. 2009;**155**:309-320

[56] Hammer GL, Dong Z, McLean G, Doherty A, Messina C, Schussler J, et al. Can changes in canopy and/or root system architecture explain historical maize yield trends in the US corn belt? Crop Science. 2009;**49**(1):299-312

[57] Bhandari M. Use of Infrared Thermal Imaging for Estimating Canopy Temperature in Wheat and Maize. Doctoral dissertation; 2016

[58] Mutawa RN. Evaluation of Sorghum Genotypes for Variation in Canopy Temperature and Drought Tolerance. Doctoral dissertation, Kansas State University; 2012

[59] Purushothaman R, Krishnamurthy L, Upadhyaya HD, Vadez V, Varshney RK. Shoot traits and their relevance in terminal drought tolerance of chickpea (Cicer arietinum L.). Field Crop Research. 2017;197:10-27

[60] Amani I, Fischer RA, Reynolds MP. Canopy temperature depression association with yield of irrigated spring wheat cultivars in a hot climate. Journal of Agronomy and Crop Science. 1996;**176**(2):119-129

[61] Mohammadi R, Armion M, Kahrizi D, Amri A. Efficiency of screening techniques for evaluating durum wheat genotypes under mild drought

conditions. International Journal of Plant Production. 2012;**4**(1):11-24

[62] Blum A. Osmotic adjustment and growth of barley genotypes under drought stress. Crop Science. 1989;**29**(1):230-233

[63] Blum A. Drought resistance, water-use efficiency, and yield potential—Are they compatible, dissonant, or mutually exclusive? Australian Journal of Agricultural Research. 2005;**56**(11):1159-1168

[64] Reynolds MP, Pfeiffer WH. Applying physiological strategies to improve yield potential. Durum wheat improvement in the Mediterranean region: New challenges. Options Méditerranéennes. 2000;**40**:95-103

[65] Blum A. Stress, strain, signaling, and adaptation–not just a matter of definition. Journal of Experimental Botany. 2015;**67**(3):562-565

[66] Hatfield JL, Gitelson AA, Schepers JS, Walthall CL. Application of spectral remote sensing for agronomic decisions. Agronomy Journal. 2008;**100**:117-131

[67] Singh D, Balota M, Isleib TG, Collakova E, Welbaum GE. Suitability of canopy temperature depression, specific leaf area, and SPAD chlorophyll reading for genotypic comparison of peanut grown in a sub-humid environment. Peanut Science. 2014;**41**(2):100-110

[68] Roohi E, Tahmasebi-Sarvestani Z, Modarres Sanavy SAM, Siosemardeh A. Association of some photosynthetic characteristics with canopy temperature in three cereal species under soil water deficit condition. Journal of Agricultural Science and Technology. 2015;**17**(5):1233-1244

[69] Pinter P, Zipoli G, Reginato R, Jackson R, Idso S, Hohman J. Canopy temperature as an indicator of differential water use and yield performance in wheat. Agricultural Water Management. 1990;**18**:35-45

[70] Pask AJD, Pietragalla J, Mullan DM, Reynolds MP. Physiological Breeding II: A Field Guide to Wheat Phenotyping. Mexico City, Mexico: CIMMYT; 2012

[71] Costa JM, Grant OM, Chaves MM. Thermography to explore plant-environment interactions. Journal of Experimental Botany. 2013;**64**:3937-3949

[72] Reynolds MP, Pask AJD, Mullan DM. Physiological Breeding I: Interdisciplinary Approaches to Improve Crop Adaptation. Mexico City, Mexico: CIMMYT; 2012

[73] Mason RE, Hays DB, Mondal S, Ibrahim AMH, Basnet BR. QTL for yield, yield components and canopy temperature depression in wheat under late sown field conditions. Euphytica. 2013;**194**:243-259

[74] Pierre S, Crossa C, Manes J, Reynolds MP. Gene action of canopy temperature in bread wheat under diverse environments. Theoretical and Applied Genetics. 2010;**120**(6):1107-1117

[75] Gutierrez M, Reynolds MP, Klatt AR. Association of water spectral indices with plant and soil water relations in contrasting wheat genotypes. Journal of Experimental Botany. 2010;**61**:3291-3303

[76] Giunta F, Motzo R, Pruneddu G. Has long-term selection for yield in durum wheat also induced changes in leaf and canopy traits? Field Crops Research. 2008;**106**:68-76

Chapter 6

Agricultural Management Strategies for Countering Drought Conditions in Eastern Croatia

Ivan Mustać, Vilim Filipović, Lana Filipović, Gabrijel Ondrašek and Dragutin Petošić

Abstract

The occurrence of drought periods which last for several months is becoming increasingly frequent, even in regions which have not encountered them before. Agricultural production is very sensitive to drought, and in areas where such conditions were rather unexpected, it is also unprepared for limited water management. As an example, in the area of the Bid-Bosut field located in eastern Croatia, a significant change in the agricultural soil water regime is noticed during a long-term study (2003–2018). From 2003 to 2018, the groundwater level at 4 m below the soil surface showed a decreasing trend of 6–10 cm annually, while this negative trend was even more prominent from 2014 to 2018 (18–71 cm annually). Furthermore, water level in a groundwater aquifer at 15 m below the soil surface showed a decreasing trend of 26–77 cm during 2015–2018. In accordance with the obtained results, this study proposes certain agro-hydrotechnical strategies which can be used in agricultural production to alleviate the effects of drought period. Although these management strategies are primarily described on an eastern continental Croatia example, they can also be applied in all agricultural areas with similar agroecological conditions.

Keywords: water availability, agricultural management, groundwater level, irrigation, monitoring

1. Introduction

Drought is commonly defined as below-usual water availability [1–3]. Even though drought is commonly associated with arid regions [4], it can also occur in more humid regions [5], which has been confirmed by climate models from various authors [6, 7]. Short-term drought periods (lasting for days or months) usually do not cause permanent or substantial environmental issues in humid areas; however, they can still be reflected on a seasonal agricultural production. Furthermore, if drought periods would last longer (for years or even decades), a negative impact on both the environment and the socioeconomic circumstances of the region can undoubtedly be expected. Agricultural crop production is particularly dependent on precipitation and therefore sensitive to the appearance of drought. Climate change predictions include uncharacteristic drought periods which besides the

IntechOpen

limited water availability for plants can also have a detrimental effect on other soil organisms (e.g., microbes) [8] and indirectly influence plant growth and development by restricting nutrient availability in soil. Thus, drought periods may lead to an overall decreased fertility of soils. Furthermore, prolonged dry conditions in soil may increase the susceptibility of soil to wind erosion, that is, frequent and severe drought periods can reduce the plant cover and expose the soil to wind, resulting in erosion and desertification [9].

Although drought periods are not unusual for the coastal parts of Croatia, in the last several decades, they have become more frequent and long-lasting, for example [10] and recently are even recorded in the continental parts of the country [11]. This is also confirmed by this hydrological study in the area of the Biđ-Bosut field, an area of 7200 ha of agricultural soil for which the construction of the Biđ field irrigation canal is planned (**Figure 1**). During the monitoring (2003–2018), significant changes of water regime, as well as soil and water management difficulties were recorded in the agricultural part of the studied area. For example, agricultural soils showed a noticeable lowering of groundwater levels, and this negative trend showed a tendency of becoming even more rapid in the future.

In the studied area, the unfavorable annual distribution of precipitation, the absence of snow cover during winter, and the rising of air temperature, all are contributing to drying conditions in agricultural soils, thus already negatively affecting local agricultural production, that is, crop rotation is becoming more and more simplified, the germination and sprouting of crops are impaired by the lack of precipitation, the yield is weather-dependent and unstable, and economic projections are frequently unreliable. Understanding of the changes in the soil water regime

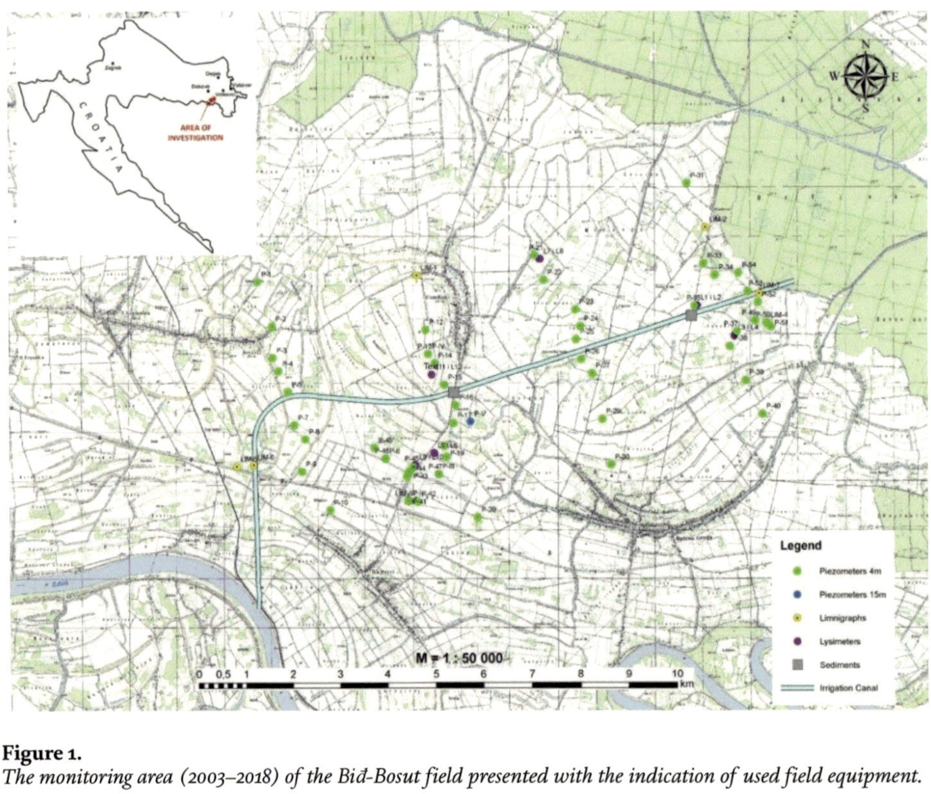

Figure 1.
The monitoring area (2003–2018) of the Biđ-Bosut field presented with the indication of used field equipment.

is of major importance for selecting the appropriate strategies for the drought risk management in agricultural systems and countering the harmful effects of climate change [12].

In this chapter, climatic and hydrological data from the Biđ-Bosut field agricultural area are presented, with a description of practical strategies which could, at least to a certain extent, alleviate the negative impact of drought on the agricultural production. One of such strategies is the installation of irrigation systems, which imposes as a relatively obvious or a simple solution, but it is not traditionally applied in agricultural production in the eastern continental Croatia, mostly because until recently drought was non-occurring or the occurrence was mostly of a relatively short duration and/or of mild intensity and because the initial cost of implementing the irrigation systems may be considered high. However, if drought periods should occur during sensitive crop developmental stages (e.g., sprouting) or extend during prolonged periods, implementing the irrigation systems could prove to be extremely beneficial for the local agriculture, as well as cost-effective in the aftermath of plant production. The fundamental basis for introducing the irrigation systems in this area was met by proceeding with the construction of the irrigation canal in 2018. However, even though the use of irrigation is a possible solution for plant production under drought conditions [13], it is by no means the only action which should be taken, especially considering that water is not an unlimited resource. Thus, the existing network of drainage canals, built in the 1960s and 1970s, could with certain modifications be used to maintain the groundwater level and contribute to the total amount of water available for irrigation. Also, the proper selection of crops and management techniques can help to facilitate plant production and, keeping the before-mentioned in mind, contribute to alleviating the negative impacts of drought on agricultural production.

2. Climate and water regime monitoring in the studied area

Climate and water regime monitoring at the Biđ-Bosut field started in 2003, with the aim of determining the impact of the irrigation canal construction on the groundwater dynamics and the surrounding agroecosystem. From the preliminary tracking of initial conditions, the monitoring evolved into a very valuable source of information with the majority of relevant agroecological data regarding the surface water, groundwater, leachate, as well as the agricultural soils in the studied area recorded. For tracking the groundwater dynamics, 50 shallow (up to 4 m of depth) and 5 deep (up to 15 m of depth) piezometers and 5 limnigraphs (up to 4 m of depth) were used. Measurements using the piezometers were done manually (by measuring tape) every 10 days, while the limnigraphs (Orpheus Mini) recorded groundwater levels daily.

The monitoring area is located at intersect between a semiarid into a semihumid moderate continental climate [14]. Meteorological data were analyzed for 2003–2018 and collected from the nearest meteorological station, the national meteorological station Gradište (45°09′ N; 18°42′ E). Data regarding the Sava river, to which the irrigation canal is connected, were taken from the Sava-Slavonski Šamac national measurement station for the years 2014–2018. The values are given on a monthly basis and are also transformed into annual values for easier following.

2.1 Precipitation dynamics

Figure 2 shows the monthly values of precipitation during the observed period. The monthly precipitation amounts (2003–2018) show a positive but nonsignificant upward trend (0.010 mm per month, that is, 0.12 mm per year), which is in accordance with the multiannual findings from similar studies [15]. In the last 5 years, a mildly negative but also nonsignificant trend (−0.40 mm per year) is visible. Moreover, the average annual sum of precipitation exhibits a mild but constant rise (682.7 mm for 1981–2018; 688.6 mm for 2003–2018; 728.8 mm for 2014–2018). Also, irregular precipitation extremes have been recorded (e.g., in June 2018; **Figure 2**). However, it should be noted here that high amounts of precipitation in a very short period actually have an extremely low effective value for crops (as explained in detail in Section 3.2).

2.2 Air temperature dynamics

The trend of increased average monthly air temperatures by 0.084°C per year (2003–2018) was recorded (**Figure 3**). If the determined value is used for the prediction of an air temperature after a longer period, data suggest that an increase of as much as 4.7°C may be expected in 50 years. The average multiannual temperature sequences also exhibit an increase (11.8°C for 1981–2018; 12.3°C for 2003–2018; 12.7°C for 2014–2018). Although these data are not sufficiently long-term in nature for solid conclusions, it is still may be considered as indicative of a general increase in air temperature.

2.3 Groundwater dynamics

Through detailed hydro-pedological research of the monitored area, a pedological characterization survey was completed with five soil-systematic units defined [16]. The classification was done according to [17], and the determined pedological units were semigleyic, hypogleyic, humogleyic, amphygleyic, and hydromeliorated soil. **Figure 4** shows the average monthly values of groundwater levels obtained by shallow piezometers located at 4 m from the soil surface, with a regard to the before-mentioned soil-systematic units.

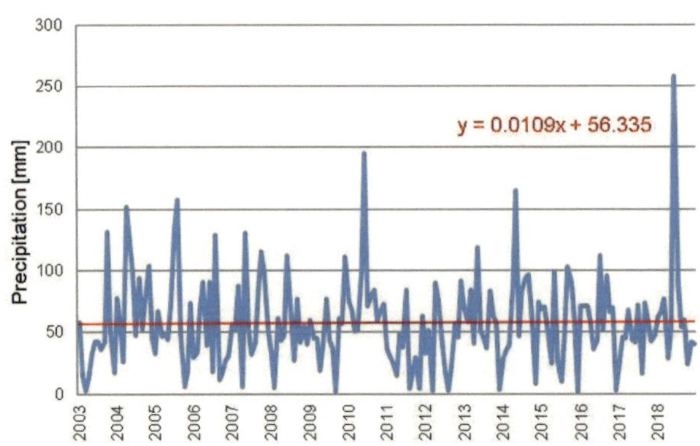

Figure 2.
The dynamics of average monthly precipitation (mm) at the monitored area of the Biđ-Bosut field.

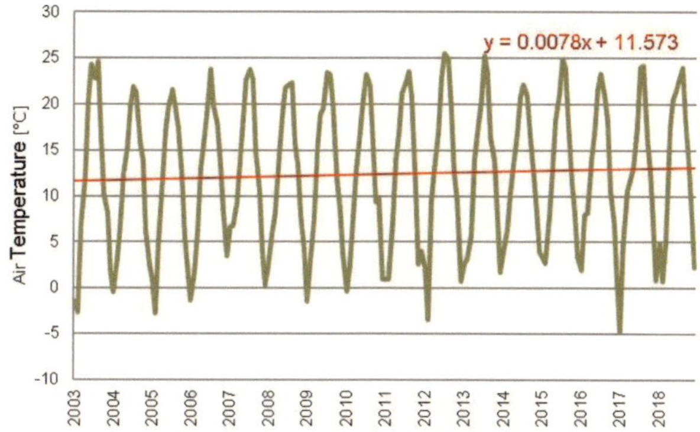

Figure 3.
The dynamics of average monthly air temperatures (°C) at the monitored area of the Biđ-Bosut field.

A relatively slight negative trend of groundwater levels (2003–2018) in the agricultural soils of the monitored area was recorded (6–10 cm per year). However, it should be noted that by observing only the last 5 years of monitoring (2014–2018), a negative trend was much more pronounce, ranging from 18 to 71 cm per year, that is, from 200% in semigleyic soil (negative trend for period 2003–2018 = 9 cm per year; period 2014–2018 = 19 cm per year) to 700% in hypogleyic soil (negative trend for period 2003–2018 = 10 cm per year; period 2014–2018 = 71 cm per year), depending on the soil type. Although the groundwater level was occasionally recorded below 4 m from the soil surface during the studied period, such occurrences were short-lasting (days) and irregular. However, in July 2017 and 2018, the groundwater level at the entire monitoring area lowered below 4 m from the soil surface and remained unchanged until the end of the year. The extremely low groundwater level which occurred in the second half of the last two research years is undoubtedly suggesting the need for further monitoring in the studied area.

The more frequent lowering of groundwater levels below 4 m from the soil surface was the reason for adding 5 deeper piezometers at 15 m from the soil surface during 2014. The average monthly groundwater level data obtained by deep piezometers are presented in **Figure 5**. The values were only slightly increased in comparison to the values measured by the shallow piezometers, which can be explained by a mild difference in pressure between the shallow soil aquifer and the deep water-bearing aquifer. The negative trend of a decreased groundwater level ranged from 26 to 77 cm per year, which is in agreement with the trends obtained by using shallow piezometers. However, data obtained from deep piezometers are for a relatively short period (4 years), and it is expected that after a longer research period, these values could even be rising.

All described climatic and water regime parameters suggest that in the studied Biđ-Bosut area, the agroecosystem changes are becoming more prominent. These changes are usually slow in progressing thus are hard to observe within shorter periods. However, field measurements and alterations of climatic and water regime parameters recorded during this study are contributing to the global predictions in which these changes in the agroecosystem are increasing in importance for the agricultural production. Finally, further continuous observation (monitoring) of climatic, water regime, and soil parameters should result in reliable databases, thus providing a foundation for the selection of appropriate site-specific strategies to counter the occurring changes and their possible negative impact on the agricultural production.

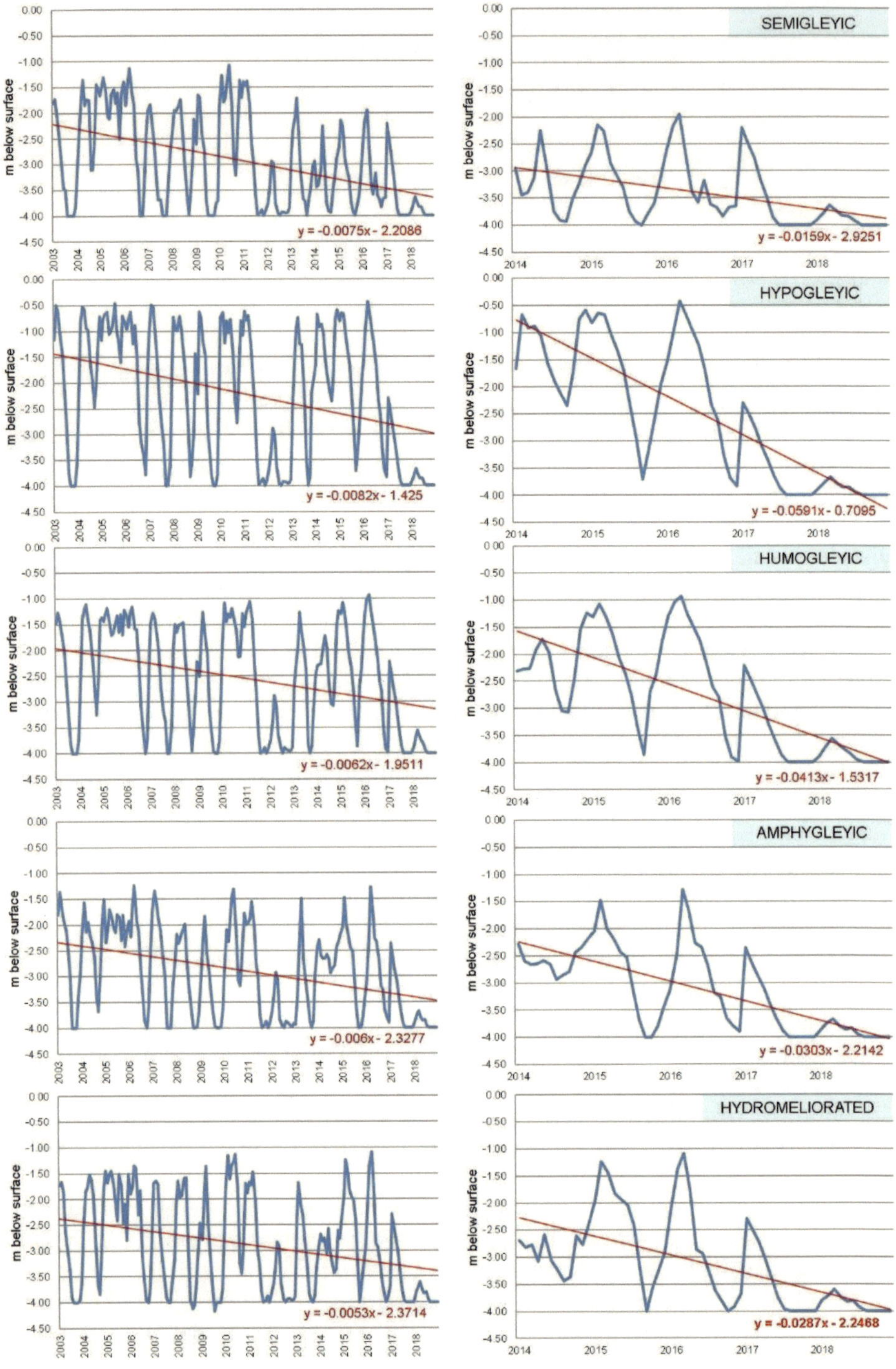

Figure 4.
The dynamics of average monthly groundwater levels in the shallow piezometers (located at 4 m from the soil surface) at the monitored area of the Biđ-Bosut field (left, groundwater levels in period 2003–2018; right, groundwater levels in period 2014–2018).

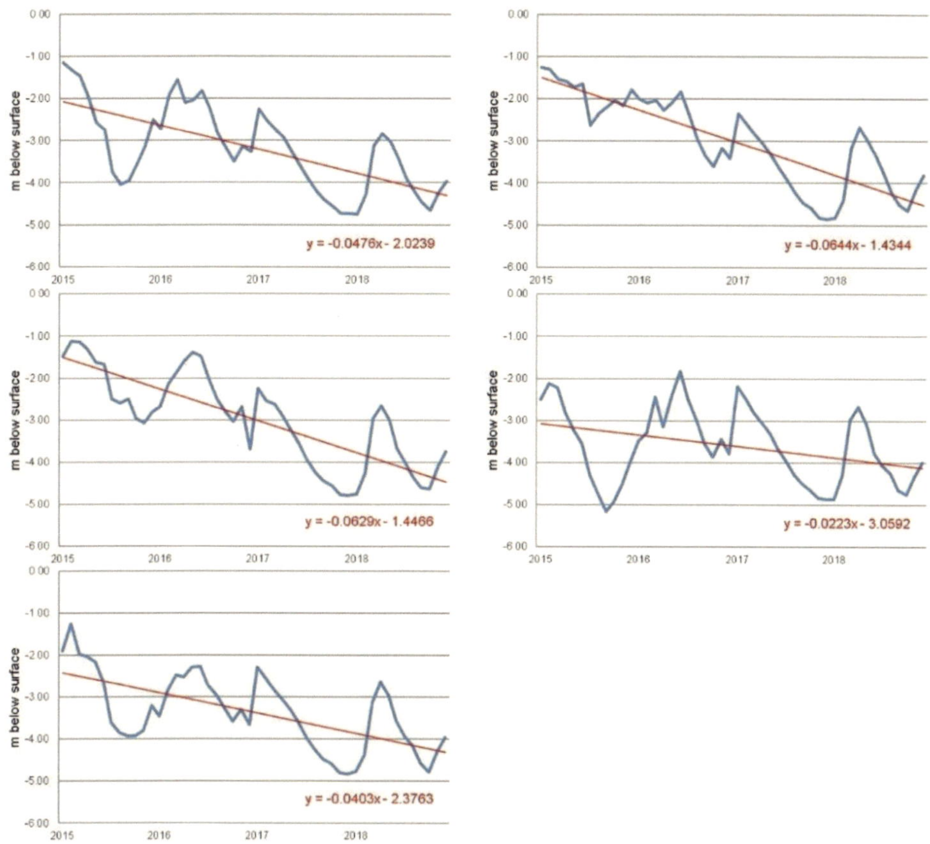

Figure 5.
The dynamics of average monthly groundwater levels in the deep piezometers (located at 15 m from the soil surface) at the monitored area of the Biđ-Bosut field.

3. Proposed measures for alleviating the consequences of drought

The measures proposed herein primarily focus not only on the eastern continental Croatia example but can also be applied to other agroecosystems with similar agroecological conditions [18]. Namely, according to the recent analyses for a 50-year period (1961–2010), it confirmed an evidence of increase in drought seasons (defined as consecutive dry days—CDD with daily precipitation <1–10 mm) notably in the eastern continental Croatia (e.g., Slavonia region) by 4%/decade to 7%/decade during summer [19]. In general, the studied agricultural area in continental Croatia characterizes relatively flat arable therein (with fluvisols, gleysols, and cambisols), positioned in between of Sava and Drava rivers, cultivated mostly with cereals and oil crops, with average annual effective precipitation of 521–890 mm and potential evapotranspiration (ET0) of 690–820 mm, as well as high irrigation demands, either in average (81–260 mm/annually) or dry (168–383 mm/annually) vegetation season [20].

3.1 Construction of the appropriate irrigation systems

Frequent periods during which groundwater level lowers below 4 m from the soil surface are imposing substantial limitations in the last decades for the agricultural production which lacks an irrigation system (such is the case in the Biđ-Bosut area). Agricultural production in the study (monitoring) area, although located in a traditionally agricultural region, so far does not rely on irrigation as a possible solution for alleviating occasional negative drought effects. The possible reason for that is because local agricultural production in this area is mostly located on hydromorphic soils, characterized by occasional or permanent moisturization by groundwater within 1 m from the soil surface [18]. Thus, the issue of lacking soil moisture which can last for several months has been an occurrence noted in this area only for the last 10 or so years, while before the main problem was the opposite: excess surface and groundwater amounts.

The completion of the irrigation canal in 2018 (**Figure 6**) was the main prerequisite for irrigation of the surrounding agricultural soils. The canal is connected to the Sava river, and, with the proper regulation of water levels in the canal, it could provide necessary and sustainable amount of irrigation water. It was projected that, during high water levels of the Sava river, water will be pumped into the melioration canal, from which it would then be channeled to the surrounding highly arable agricultural fields during the most of vegetation season given on negative water balance, for example [20]. More precisely, considering the amount of water in the Sava river [18], this hydrotechnical solution could help to ensure adequate amount of water for irrigation of the approximately 10,000 ha of surrounding agricultural soil. As for the quality of the water, studies from various authors have made it clear that the water from the Sava river is of ample quality to irrigate the local crops, for example [18]. However, water quality is an important factor when considering its use for crop irrigation; thus, if canal water is used for irrigation, it is necessary to implement permanent water quality control.

Application of appropriate water management strategies for the usage of Sava river water for irrigation of crops is of major importance. Such strategies include application of the modern low-pressurized/low-energized (fert)irrigation system, adaptation of cropping pattern (e.g., to give advantage to winter over spring cereals/cultivars and to those with shorter vegetation period when water balance is the most negative), modernization of conveyance systems (e.g., channel overlying or replacing with pipelines), conduct irrigation management on real time data measurements, application of conservation agriculture practices, and many others [20, 21]. Some of the most recent studies have confirmed that almost all crops cultivated on the studied areas are exposed to water stress (negative water balance) with significant yield losses even in normal (average) sessions. For instance, in Brodsko-posavska County (overspread on the most of elaborated area), an average annual (for 1963–2005 period) effective precipitation reaches 690 mm, while potential evapotranspiration (ET0) is 718 mm, causing the negative water balance during vegetation period for almost −200 mm [22]. According to the same study, irrigation requirements in average climate season for the most cropped cultures yield from 82 mm (corn) up to 160 mm (sugar beet) and over 200 mm (lucerne), while in dry seasons water requirements are higher by 1.8–1.9-fold (lucerne and sugar beet) up

Figure 6.
Opening of the Biđ-Bosut field irrigation canal.

to 2.6-fold (corn). The yield reduction in the case of nonirrigated conditions on this area is also significant for the most of crops, even in normal (from 11% in corn and soybean up to 25% in sugar beet on texture-lighter soils) and especially in dry seasons (from 25% in soybean up to 47% in sugar beet on texture-lighter soils).

Although Sava river can provide the required amount of water for irrigation, excessive (unsustainable) management measures could possibly create additional (agro)ecological issues regarding water levels of Sava river and even question the sustainability of such practice. This possibility is confirmed by the trend of lower levels of Sava river by 0.51 m per year for the period from 2014 to 2018 (**Figure 7**). Using Sava river water for irrigation should therefore be applied with the utmost rationality, that is, taking into consideration the optimal water regime within the river-soil-plant-atmosphere system, for example [23]. Additionally, the education of local farmers should be included as an important step in the planning and implementation of any irrigation system which is depending on a natural system, such as (Sava) river.

3.2 Using the existing irrigation infrastructure for the purpose of collecting precipitation

During the 1960s and 1970s, the main issue for the agricultural production in the studied region was the excess amount of surface and groundwater. That is why the area has an abundance of drainage canal networks through which excessive water was channeled into the recipient—the Sava river.

However, in June 2018, 257.4 mm of precipitation was recorded (**Figure 2**), which exceeds the average monthly precipitation in this area by several times (320%). These extremes were usually accompanied by storms and hail, which is why the authorities declared a state of emergency for the years 2010, 2014, and 2018. As mentioned before, such high amount of precipitation in a very short time has a very small effective value for crops because in such conditions water cannot infiltrate in the soil, usually resulting in (sub) surface runoff. In the studied area, most of the water from surface runoff streams firstly into the drainage canals, then toward the Sava river, and finally reaching the Danube river and ultimately the Black Sea. Thus, water from surface runoff is basically lost from this area and does not have any effect on the water regime of the soils, although the possibilities and sustainability of some on-farm water storage systems (e.g., surface accumulations, public reservoirs) should be also evaluated. This was confirmed by field measurements (**Figures 2-5**), from which it is clear that even the abundant rainfall in June

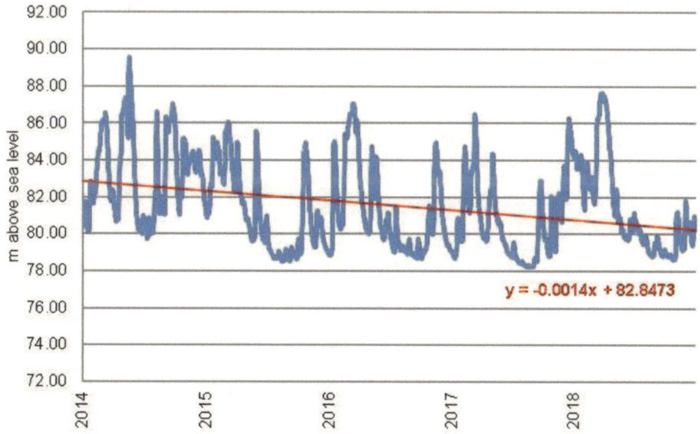

Figure 7.
The dynamics of Sava river daily levels at the Sava-Slavonski Šamac measuring station for the period from 2014 to 2018.

2018 did not lead to a noticeable rise in groundwater levels. Moreover, for the whole first half of 2018, the groundwater did not exceed the level of 3.7 m from the soil surface, and in July of the same year, groundwater level lowered below 4 m from the soil surface, remaining at stated level until the end of the year 2018.

The network of drainage canals was up until 10 years ago used exclusively for drainage of the excess water from the area. During the last 10 or so years, the appearance of excess water became increasingly rare, and in 2017 and 2018, no such occurrence was recorded, except for a few days in June 2018 (data not shown). What is more, in the last several years, the lack of moisture in soil has become especially noticeable and culminated in 2018. One possible hydrotechnical solution for such issue would be to modify the existing canal network by implementing the controlled drainage canal system (where water flow is controlled and limited by a regulating system) at the main drainage canals. This way, in cases of an extreme precipitation, the drainage canals would preserve their primary drainage function, but in case of lower precipitation (when no excessive water is present in soil but before the drought conditions), by closing the canal release point, the same canals could be used as a form of a precipitation retention system. This proposed system of a branched-out canal network could, with an adequate regulation of canal water release points, prove to be very useful when additional amount of water for the agricultural plant production is necessary, that is, under drought conditions. Using these drainage waters as a potentially valuable "resource" rather than considering them as a "waste" can contribute to the alleviation of water scarcity, thus the negative effects of drought conditions [24], which is also in accordance with the widely accepted and nowadays preferential concept of sustainability in agricultural production.

Additionally, if subsurface drainage systems are installed, there is also the possibility of implementing the subsurface drainage water regulation system which could control the groundwater level according to the soil moisture. According to [25], controlled drainage, also known as drainage water management, is a practice of using the water control structure at the drainage outlet in order to raise the groundwater level and thus retain water in soil during periods when drainage is not needed, but a deficiency of soil water is present. The implementation of controlled systems ($120 or $50–100 per ha if upgrading from conventional drainage systems) is relatively inexpensive [26] and therefore should be taken into account when designing an agricultural systems. However, considering the initial cost of installing such system, its introduction should be accompanied by a sufficiently profitable agricultural production that would presumably justify the additional investment.

3.3 Selection of crops and growing techniques in agricultural areas without the irrigation systems

Agricultural production without an irrigation system is completely depending on climate and available soil moisture (weather-dependent). In the context of increasingly important climate change, such production will presumably encounter more and more stressful conditions (i.e., plant water stress). In order to maintain the productivity, drought- and heat-tolerant crops/cultivars/hybrids must become the product of choice, as must the application of techniques to maintain the soil moisture by reducing evaporation [27, 28]. More precisely, evaporation occurs when moist soil is exposed to the atmosphere. In theory, to reduce the evaporation, it is necessary to reduce the exposed soil surface as much as possible and/or to shorten the time of the soil exposure to the atmosphere. In practice, mulching with plant residues and/or polyethylene foils can be used for this purpose [29]. Also, certain probiotic soil enhancers which have become available on the market recently can be used for the same purpose of reducing the evaporation [30]. These soil enhancers enrich the soil with beneficial microbes which accelerate decomposition of soil organic matter into smaller compounds capable to retain more water in the soil and further to plant-available nutrients, which increases the overall soil fertility but also improves soil capacity to retain moisture. Additionally, if the irrigation systems are applied in the studied area of Biđ-Bosut field, the appropriate irrigation systems are those with the localized water distribution (e.g., micro-sprinklers or drip irrigation), which distribute water only alongside the crops and thus reduce water losses and evaporation (in comparison with, e.g., irrigation boom).

At the end, the important viewpoint of the drought-alleviating management techniques is also from the economical aspect. Generally, adequate agricultural management includes the cost-benefit ratio regarding the crop value. Higher input into the agricultural production should be justified by investing into profitable crops, which will presumably pay out the initial investment. In this context, replacing the less profitable crops with crops for which the market demand is higher could be an appropriate action. However, this agricultural management strategy is not an easy task as it is not grounded on a permanent aspect but strongly relies on the current supply and demand market circumstances. Thus, additional economic analyses which include supplementary perspectives such as estimations of future market opportunities and trends may be of major importance.

4. Conclusion

Climatic and soil water regime data (2003–2018) suggest that the agroecosystem changes are becoming more prominent in the studied Biđ-Bosut area, and thus the future agricultural production may be exposed to the greater pressures regarding the insufficient amount of water in the soil. Also, some of the most recent midterm climate scenarios (models) performed for the studied and wider area support our theses. For instance, modern climate models from local to global scales employ relatively different horizontal resolutions from 10 to 300 km [19] and predict wide range of climate parameters, that is, scenarios. At the European scale (notably in its central part), it is expected that average seasonal near-surface temperature (Ta) is going to increase in the period 2011–2040 by 0.2–2°C [19]. According to the same authors and for the same midterm period, in Croatia the largest changes in Ta can be expected in the mid of vegetation session (summer) with an increase of Ta by 0.8–1°C in the central part of Croatia and around 0.8°C in eastern (Slavonia) region. As regards the average precipitation, a decrease of precipitation between 2 and 8% is predicted over the larger part of Croatia [19]. Consequently, higher evapotranspiration demands (over increasing average vegetation air temperature) and reduced average effective precipitations might further exacerbate water imbalances in the agroecosystems on the elaborated area.

Installation of the irrigation systems is a possible solution for countering the negative impact of drought, but other management strategies should also be implemented in order to achieve the sustainability of agricultural production. In this context, the education of local farmers should be included as an important step in the planning and implementation of any drought countering techniques, in order to achieve the highest success rate by adhering the rules and instructions referring to the rational and responsible water use. Finally, this study has shown that multiannual climate and soil water regime data may provide a good basis for the decision-making process in creating sustainable agricultural management policies (construction of the appropriate irrigation systems and use of the existing irrigation infrastructure for the purpose of collecting precipitation, use of drought- and heat-tolerant crops/cultivars/hybrids, and application of techniques to maintain the soil moisture by reducing evaporation, for example, mulching with plant residues and/or polyethylene foils and use of probiotic soil enhancers) focused on countering the negative impact of drought on the agricultural production.

Author details

Ivan Mustać*, Vilim Filipović, Lana Filipović, Gabrijel Ondrašek and Dragutin Petošić
Faculty of Agriculture, University of Zagreb, Zagreb, Croatia

*Address all correspondence to: imustac@agr.hr

IntechOpen

© 2019 The Author(s). Licensee IntechOpen. This chapter is distributed under the terms of the Creative Commons Attribution License (http://creativecommons.org/licenses/by/3.0), which permits unrestricted use, distribution, and reproduction in any medium, provided the original work is properly cited.

References

[1] Wilhite DA, Glantz MH. Understanding: The drought phenomenon: The role of definitions. Water International. 1985;**10**(3):111-120. DOI: 10.1080/02508068508686328

[2] Tallaksen LM, Van Lanen HAJ, editors. Hydrological drought: Processes and estimation methods for streamflow and groundwater. In: Developments in Water Science. Vol. 48. Amsterdam, The Netherlands: Elsevier Science B.V; 2004. ISBN 978-0-444-51767-8

[3] Mishra AK, Singh VP. A review of drought concepts. Journal of Hydrology. 2010;**391**(1-2):202-216. DOI: 10.1016/j.jhydrol.2010.07.012

[4] Güneralp B, Güneralp İ, Liu Y. Changing global patterns of urban exposure to flood and drought hazards. Global Environmental Change. 2015;**31**:217-225. DOI: 10.1016/j.gloenvcha.-2015.01.002

[5] Goddard S, Harms SK, Reichenbach SE, Tadesse T, Waltman WJ. Geospatial decision support for drought risk management. Communications of the ACM. 2003;**46**(1):35-37. DOI: 10.1145/602421.-602442

[6] Sheffield J, Wood EF. Projected changes in drought occurrence under future global warming from multi-model, multi-scenario, IPCC AR4 simulations. Climate Dynamics. 2007;**31**(1):79-105. DOI: 10.1007/s00382-007-0340-z

[7] Seneviratne SI, Nicholls N, Easterling D, Goodess CM, Kanae S, Kossin J, et al. Changes in climate extremes and their impacts on the natural physical environment. Managing the Risks of Extreme Events and Disasters to Advance Climate Change Adaptation. Special Report of the Intergovernmental Panel on Climate Change. 2012:109-230. DOI: 10.1017/cbo9781139177245.006

[8] Gschwendtner S, Tejedor J, Bimueller C, Dannenmann M, Knabner IK, Schloter M. Climate change induces shifts in abundance and activity pattern of bacteria and archaea catalyzing major transformation steps in nitrogen turnover in a soil from a mid-European beech forest. PLoS One. 2014;**9**(12):e114278. DOI: 10.1371/journal.pone.0114278

[9] Jones A, Panagos P, Barcelo S, Bouraoui F, Bosco C, Dewitte O, et al. The State of Soil in Europe. Luxembourg: Publications Office of the European Union. Available from: http://publications.jrc.ec.europa.eu/repository/bitstream/JRC68418/lbna25186enn.pdf; 2011 [Accessed: 27 March 2019]

[10] Romić D, Romić M, Ondrašek G, Zovko M, Bubalo M, Bakić H, et al. Monitoring zaslanjenja voda i poljoprivrednih tala na području doline Neretve. 2016; (report)

[11] Mustać I, Petošić D, Filipović V, Tomić F, Ondrašek G, Igrc MD, et al. Monitoring vodnog režima poljoprivrednih tala i kakvoće vode na području Dovodnog melioracijskog kanala za navodnjavanje Biđ-bosutskog polja za 2018. godinu. 2019; (report)

[12] FAO and ITPS. Status of the World's Soil Resources (SWSR)—Main Report. Rome, Italy: Food and Agriculture Organization of the United Nations and Intergovernmental Technical Panel on Soils; 2015. ISBN 978-92-5-109004-6

[13] Kour G, Bakshi P. Irrigation management practices and their influence on fruit agroecosystem. In: Ondrašek G, editor. Irrigation in Agroecosystems. Intech Open; 2019. DOI: 10.5772/intechopen.79569

[14] Köppen W. The thermal zones of the earth according to the duration of hot, moderate and cold periods and to the impact of heat on the organic world. Meteorologische Zeitschrift. 1884;**20**(3):351-360. (published 2011). Translated by Volken E; Brönnimann S

[15] Perčec Tadić M, Gajić-Čapka M, Zaninović K, Cindrić K. Drought vulnerability in Croatia. Agriculturae Conspectus Scientificus. 2014;**79**(1):31-38. Available from: https://hrcak.srce.hr/120753 [Accessed: 27 March 2019]

[16] Petošić D, Mustać I. Značajke vodnog režima i kakvoće poljoprivrednih tala na području Višenamjenskog kanala Dunav-Sava. Zagreb: Monografije Sveučilišta u Zagrebu-Agronomski fakultet. 2010. ISBN 978-953-6135-92-9

[17] Husnjak S. Sistematika tala Hrvatske. Udžbenici Sveučilišta u Zagrebu; 2014. ISBN 978-953169267-0

[18] Romić D, Marušić J, Tomić F, Petošić D, Šimunić I, Romić M, et al. Nacionalni projekt navodnjavanja i gospodarenja poljoprivrednim zemljištem i vodama u Republici Hrvatskoj. Agronomski fakultet Sveučilišta u Zagrebu; 2005

[19] Branković Č et al. Sixth National Communication of the Republic of Croatia under the United Nation Framework Convention on the Climate Change (UNFCCC) Selected Sections in Chapters: 7. Climate Change Impacts and daptation easures 8.Resarch, ystematic bservation and Monitoring. Zagreb: Meteorological and hydrological service of Croatia; 2013

[20] Ondrasek G. Water scarcity and water stress in agriculture. In: Ahmad P, Wani M, editors. Physiological Mechanisms and Adaptation Strategies in Plants Under Changing Environment. New York, NY: Springer; 2014. DOI: 10.1007/978-1-4614-8591-9_4

[21] Ondrašek G. Irrigation in agroecosystems. In: Ondrašek G, editor. Introductory Chapter: Irrigation after Millennia - Still One of the Most Effective Strategies for Sustainable Management of Water Footprint in Agricultural Crops. London, United Kingdom: IntechOpen; 2019. DOI: 10.5772/intechopen.73607

[22] Brodsko-posavska County (BPC). Conceptual solution of irrigation system Slobodnica. 2015

[23] Ondrašek G, Rengel Z, Petošic D, Filipović V. Land and water management strategies for the improvement of crop production. In: Ahmad P, Rasool S, editors. Emerging Technologies and Management of Crop Stress Tolerance. London, United Kingdom: Academic Press; 2014. pp. 291-313. ISBN 9780128008751

[24] Shideed K, Mazid A, Oweis T, van Ginkel M. Potential of rainfed agriculture and smallholder farmers in food self-sufficiency. In: Sadik A, El-Solh M, Saab N, editors. Arab Environment: Food Security. Annual Report of the Arab Forum for Environment and Development, Chapter: Potential of Rainfed Agriculture and Smallholder Farmers in Food Self-Sufficiency. Beirut, Lebanon: Technical Publications; 2014. pp. 74-102. Available from: https://www.researchgate.net/publication/313846915_Potential_of_Rainfed_Agriculture_and_Smallholder_Farmers_in_Food_Self-Sufficiency [Accessed: 27 March 2019]

[25] Fausey NR, Brown LC, Belcher HW, Kanwar RS. Drainage and water quality in Great Lakes and Cornbelt States. Journal of Irrigation and Drainage Engineering.

1995;**121**(4):283-288. DOI: 10.1061/(asce)0733-9437(1995)121:4(283)

[26] Richard C, Gary R, Larry B. Drainage Water Management: A Practice for Reducing Nitrate Loads from Subsurface Drainage Systems. Washington, DC: United States Environmental Protection Agency (EPA); 2006

[27] Veres S, Ondrašek G, Zsombik L. Wheat sensitivity to nitrogen supply under different climatic conditions. In: Fahad S, Basir A, Adnan M, editors. Global Wheat Production. London, United Kingdom: Intech Open; 2018. DOI: 10.5772/intechopen.76154

[28] Solh M, van Ginkel M. Drought preparedness and drought mitigation in the developing world's drylands. Weather and Climate Extremes. 2014;**3**:62-66. DOI: 10.1016/j.wace.2014.03.003

[29] Filipović V, Romić D, Romić M, Borošić J, Filipović L, Mallmann FJK, et al. Plastic mulch and nitrogen fertigation in growing vegetables modify soil temperature, water and nitrate dynamics: Experimental results and a modeling study. Agricultural Water Management. 2016;**176**:100-110. DOI: 10.1016/j.agwat.2016.04.020

[30] Pathak RK, Ram RA. Bio-enhancers: A potential tool to improve soil fertility, plant health in organic production of horticultural crops. Progressive Horticulture. 2013;**45**(2):237-254. Available from: http://krishikosh.egranth.ac.in/handle/1/5810021821 [Accessed: 27 March 2019]

Chapter 7

Drought Stress Response in Agricultural Plants: A Case Study of Common Bean (*Phaseolus vulgaris* L.)

Aleš Sedlar, Marjetka Kidrič, Jelka Šuštar-Vozlič, Barbara Pipan, Tanja Zadražnik and Vladimir Meglič

Abstract

Drought is one of the major threats to common bean (*Phaseolus vulgaris* L.), affecting its growth and productivity and, thus, contributing to considerable losses in yield in many regions worldwide. The development of varieties tolerant of drought stress has, therefore, become one of the primary goals in many common bean breeding programs. Plants have developed various mechanisms of their adaptation, to a greater or lesser extent, to drought. These are expressed, on the molecular level, by changes of gene expression and of protein content, together with responses at physiological and morphological levels. The response of common bean to drought is still not sufficiently well characterized due to its genetic complexity and its diverse, often ambiguous, phenotypic effects. Understanding these mechanisms is thus of fundamental importance for developing varieties that are better adapted to such stress conditions. In this chapter, we present research that provides an insight into the morpho-physiological adaptation and its underlying molecular changes in common bean plants exposed to drought. We include our contribution to establishing the basis for breeding of common bean with greater tolerance to this abiotic stress that uses molecular markers and identification of quantitative trait loci (QTLs).

Keywords: common bean, drought, transcriptomics, proteomics, posttranslational modifications, quantitative trait loci (QTLs), breeding

1. Introduction

Common bean (*Phaseolus vulgaris* L.) is cultivated worldwide and accounts for one-third of cultivated legumes (FAOSTAT, 2014). The biggest producers are Brazil, USA and Mexico, while millions of tons are produced annually in Africa. Its high levels of protein and carbohydrate make common bean a staple in food, and it is an important source of vitamins and minerals, including iron, magnesium, potassium, zinc, selenium and calcium [1–4].

It requires a considerable amount of water during its growing season in order to develop and produce seeds in accordance with its genetic potential [1, 5]. Exposure to drought can thus result in diminution of fitness, quality and quantity of yield,

IntechOpen

depending on the timing, intensity and duration of drought, as well as on the stage of development of the exposed plant. Exposure to drought stress can be fairly constant throughout the season or may affect plants in a specific stage of their life cycle, thus delaying early plant development, vegetative growth, flowering and/or maturation [6]. In the most arid areas, annual losses of common bean yield can exceed 60%, rising to 80% at the height of the drought [5–7].

Plants have, during evolution, acquired different modes of adaptation to harsh environmental conditions, including drought stress [5, 8, 9]. Some plants escape the latter by early maturation and, in consequence, more rapid development of seeds, thus completing their life cycle before the onset of prolonged and severe drought. Others have developed morphological, anatomical or physiological adaptations that enable them to maintain high water potential during drought (drought avoidance), or mechanisms directed to survival in the presence of low water potential (drought tolerance). Adaptations leading to drought avoidance include increased density of roots and their deeper propagation that enables more effective water absorption from the soil, and decreased leaf area and stomata closure that control the limitation of water loss through transpiration [10]. Drought tolerance is based on tissue- and cell-specific physiological and molecular adaptations such as synthesis of osmoprotective proteins like dehydrins and chaperons [11–13].

An outstanding characteristic of common bean germplasm is its particularly high diversity. In the second part of this chapter, the general picture obtained by commonly used screening methodology is presented. This includes phenotyping of phenological, anatomical and morphological, as well as yield associated traits, illustrating this diversity with the aim of pointing out that common bean responses to drought can differ greatly between specific genotypes. An example of this approach is that phenotyping of American genotypes in the last decades led to the discovery of important drought resistance sources from the Mesoamerican gene pool, largely belonging to the race Durango, thus forming the basis for numerous subsequent studies [5]. For such reasons, continuous efforts are devoted to screening common bean germplasm for more drought resistant genotypes that exhibit different, and potentially complementary, drought resistance traits that are and will be used to study and better understand the mechanisms of resistance and for the breeding of new varieties.

The plant response to drought results from complex and diverse adaptation. It has, therefore, to be studied on levels ranging from morphological and physiological changes observed in organs to the intricate responses on the gene expression and regulation levels and to biochemical responses on the level of cells and organelles [12]. In the present chapter we therefore focus further on the physiology of the response of common bean to drought, followed by a survey of research on the influence of drought on its transcriptome, proteome and post-translational modifications.

In the last part of this chapter the genetic level is considered, disclosing drought response as a complex quantitative trait controlled by a number of major and minor genes clustered on specific loci, as well as several genomics and molecular approaches that have been utilized for their study [14]. For common bean genotyping and for subsequent mapping of quantitative trait loci (QTLs) a large variety of genetic markers, from simple sequence repeat (SSR) to single-nucleotide polymorphism (SNP) markers, are now available, enabling use of this approach in identifying molecular markers of tolerance to drought.

Further, identification of genes that are expressed differently under drought conditions from that in well-watered plants, especially if contrasting tolerant and sensitive genotypes, may also lead to the discovery of specific markers that can be used in breeding. Similarly, protein markers can be discovered by proteome profiling studies by identification of differentially expressed proteins in regard to the drought conditions or genotype sensitivity to drought [15–17]. All further attempts at successful breeding of common beans with improved tolerance to drought will benefit greatly from the determination of the genome sequence of common bean that has enabled the development and identification of novel molecular markers and completion of a comprehensive common bean consensus map applicable to both Andean and Mesoamerican genotypes [18].

2. The diverse common bean germplasm as a potential for discovering new drought-responsive traits

Common bean originates from Mexico and was separated in two ecologically and geographically different Mesoamerican and Andean gene pools [19, 20]. Mesoamerican bean genotypes can be distinguished by longer flowering times and small seeds, while Andean genotypes have large and colorful seeds [21, 22]. Phylogenetic studies and evaluation of common bean genotypes collected from different regions, ranging from the Americas, Africa [23, 24] and Europe [25–28], have confirmed independent domestication events specific to each of the gene-pools [20, 29].

Common bean was introduced in Europe centuries ago by independent domestication events from both major centers of origin [25–28]. Our phylogenetic studies shed more light on the understanding of dissemination pathways and the evolution of this species in central Europe and have been focused on the germplasm from the Central European, South East European and Balkan region [27, 28, 30]. Evaluations of genetic diversity and the population structure of 167 historical and current accessions with the different geographical origin (Slovenia, Austria, Bosnia and Herzegovina, Croatia, Macedonia and Serbia) have revealed great allelic polymorphism in 14 SSR markers. The strong predominance of Andean genotypes in Slovenia and several Western Balkan countries indicates their introduction from the Mediterranean basin and countries such as Italy. On the other hand, a high proportion of Mesoamerican genotypes in the present Austrian germplasm (44%) could indicate their introgression from western and northern European countries driven by historical events.

Cultivation in diverse local environments and climate areas, ranging from lowlands to high altitude regions in equatorial and more temperate climate conditions have contributed to the high diversity of common bean in terms of growth type, seed properties and maturity time [31]. Consequently, the diverse common bean germplasm represents an important trait pool for searching for new traits such as abiotic stress tolerance traits [31]. The screening process is largely based on phenotyping genotypes exposed to a form of drought, with possible subsequent rehydration in comparison to irrigated conditions. It is usually performed in different locations, over several seasons, either in the field or in controlled greenhouse conditions. Screening commonly includes phenotyping of phenological traits such timing of flowering and maturation, anatomical and morphological traits describing plant fitness and yield-associated traits as a measure of the effect of drought on yield output.

In the past decades, considerable efforts have been made to characterize the American germplasm for different traits including drought responsive traits [5]. The identified drought resistance sources of race Durango have become a cornerstone for research of complex drought tolerance mechanism and introgression of traits into the cultivars.

Another example is Central and East European common bean germplasm that consists of thousands of collected genotypes deposited in the national and regional gene banks and preserving this variability is an important step in preventing gene erosion, as well as supporting breeding programs with genotypes showing different environmental adaptations. Characterization and evaluation of this germplasm are an ongoing process and have confirmed the very broad genetic diversity of common bean in Eastern Europe. Our recent proceedings have resulted in formation of a core collection having applicative value for direct breeding purposes [32]. Screening for representative genotypes for core collection included initial evaluation of basic multi-crop passport descriptors (e.g., geographic origin, biological status, and ancestral data), phenotypic seed characteristics and phaseolin type, as well as assessment of genetic structure by genotyping with genetic markers. The resulting core collection encompasses 63 accessions representing the global genetic diversity and 14 standard genotypes with desirable traits from the East European region (unpublished data) and was evaluated under field conditions as well as for the presence of genetic markers associated with traits of interest and biochemical analysis. Core collection was further evaluated for agronomic traits in field conditions (response to abiotic stress), genetic markers for desirable traits and nutritional traits of importance (multi-elemental composition, fats, proteins, and phytic acid). These results enabled selection of superior genotypes in core collection for further breeding applications.

3. The physiology of drought response

The observed physiological changes in plants exposed to drought can be a direct consequence of drought, as well as of the response of the plant, in order to mitigate the stress. Drought typically occurs as a result of low and non-frequent precipitation, resulting in reduced soil water content that is first detected by plant roots [33]. Depending on drought duration and severity the water status of plants can be affected by insufficient water absorption due to low soil water availability, as well as to increased water loss in the process of transpiration. Water deficit in plants affects their normal physiological processes and hinders the development, growth and yield, ultimately resulting in wilting, senescence and plant death [33]. Drought can be potentiated by heat stress, which propagates the water loss from the plant by increased water evaporation from the leaves, and by soil salinity stress, together reducing soil water availability as well as having an additional toxic effect on the plant [34–36].

Drought responsive traits have been studied in common bean by evaluating various traits in field experiments as well as in more controlled environments, such as, greenhouses [37]. Phenological and yield-associated traits have been studied frequently because they are affected by drought stress, are an important indicator of yield output and are also measurable in a large phenotyping population. Phenological traits can also represent the adaptation of the life cycle of a plant to specific drought conditions in the environment. Breeding for earliness is an effective strategy for increasing the yield stability in regions such as the Mediterranean, where plants are exposed to increased drought in the summer time. Early flowering can help in drought avoidance; however, it is not effective in mitigating the drought stress once it occurs [9].

The response to drought in different common bean genotypes has been characterized in several studies based on physiological measurements such as photosynthesis and photosynthate acquisition as well as on partitioning indices [5, 38–44]. Photosynthesis and cell growth are primary processes influenced by drought due to decreased stomata conductivity in the early drought phases that limits evaporation and CO_2 diffusion in the leaf mesophyll. The surplus of energy on the thylakoids in the photosynthesis apparatus results in photo inhibition—reduced photochemical efficiency [45]. Stomata closure is an effective strategy for shorter drought periods and for mild drought where photosynthesis is not affected in such a way as to reduce the yields [38]. When the duration of drought is longer, better drought tolerance and yields are enabled by a specific biochemical mechanism on the cell level. High yields of tolerant Durango genotypes, such as, 'Pinto Saltillo,' exposed to drought have been associated with early and fine regulation of the stomata response and with CO_2 assimilation with stomata closure, limiting water loss during the day, maintaining higher relative water content (RWC) at night, with increased water use efficiency and limitation of reactive oxygen species (ROS) accumulation [41, 42, 44].

Screening selected genotypes adapted to Central European climatic conditions enabled us to identify genotypes with more drought tolerant traits [46, 47] thus becoming a starting point for studies on the mechanisms of drought. They were performed under controlled conditions with drought being induced by discontinuing irrigation and assessment of drought by soil water potential measurements. Observation of plant physiological changes, measurements of leaf water potential, relative water content and yield were employed for determining drought tolerance. Among the tested cultivars adapted to the growing conditions of the Central European region, the greatest difference in response to drought was observed between 'Tiber' and 'Starozagorski Čern,' the former being the most tolerant. For this reason these two cultivars were used in many of our studies of the response to drought [46, 47]. Studies on the level of physiology have confirmed that the ability to withstand drought is also related to the water consumption pattern of the plant. Some cultivars such as 'Starozagorski Čern' exhibit water spending behavior that enables them to thrive, and, when the water supply is sufficient, they produce high yields [47]. However, when exposed to drought, their yield can be significantly reduced. In contrast, cultivars adapted to harsher environments, regulate water more conservatively and their yield during drought is affected less, as is the case with 'Tiber.' Drought tolerance in water saving cultivars has been associated with great plasticity on the biochemical and cellular levels, being associated with stomatal conductance, photosynthesis rate, abscisic acid (ABA) synthesis and resistance to photoinhibition [39]. In addition, the distribution of photosynthetic products to developing pods and seeds is an important factor in determining the yield under stress, with genotypes with better partitioning indices expressing a higher yield [5, 40, 43].

4. Response to drought on the molecular level

Understanding the changes in metabolic pathways in plants under the influence of drought, as well as the molecular mechanisms regulating their adaptation to this stress, is very important in identifying key molecular markers that could help distinguish between genotypes with different tolerance. On the molecular level, drought affects plant cells in different ways—through changes in gene expression and/or translation of transcripts to proteins, through posttranslational modification leading to protein activation and by further direct action on the protein itself. It is important to underline that these ways are interdependent and that only active key proteins enable a response beneficial for the plant.

Screening genes with differential expression and proteins with changed abundance or activity in plants exposed to drought is greatly facilitated by modern transcriptomic and proteomic tools which have, together with other approaches, enabled rapid development of the field. Identification of detected genes and proteins is greatly facilitated by the recent advances in sequencing and the publication of full genome sequences of model legumes *Medicago truncatula* [48] and *Lotus japonicus* [49] and of crop legumes, such as, common bean [50], soybean [51], chickpea (*Cicer arietinum*) [52] and peanut [53]. Comprehensive lists of genes and proteins obtained from screening studies are then classified according to their known ontologies in order to further investigate their interactions and connections, by methods of bioinformatics and systems biology, into metabolic pathways. A combination of these approaches has enabled the identification of thousands of genes with differential expression and hundreds of proteins with changed abundance in common bean under drought [15, 17, 54–56] as well as other model, crop and forage legumes [57].

4.1 Transcriptomic profiling of drought response

Early transcriptomic profiling methods employed over the past decades utilized polymerase chain reaction (PCR) and hybridization techniques and allowed for detection of a smaller number of transcripts with a large difference in mRNA abundance between compared samples [58, 59]. Today these methods are being replaced by genome-wide profiling techniques, such as, microarrays and whole-genome mRNA sequencing (RNA-seq). Further intricacies of gene regulation are explored by profiling miRNAs, small non-coding RNAs that regulate gene expression [55, 56, 60–62].

The transcriptomic response of common bean has been investigated in various plant organs of different genotypes with respect to different stages of drought severity [56]. One of the first studies focused on roots, the first plant organ in which to detect changes in soil water content [55]. Several dehydration-related genes were identified that are associated with signaling, protein homeostasis and root growth modulations, among which a gene *PvOCT1*, encoding a new type of organic cation transporter in plants, has been reported [63]. The response in leaves is equally important since the regulation of transpiration plays an important role in the plant response to drought. We showed that in leaves of eight common bean genotypes at different levels of dehydration, up-regulation of transcription factors and genes encoding osmoprotectants, late embryogenesis abundant (LEA) proteins, protein kinases, aldehyde dehydrogenases and cell and carbohydrate metabolism-associated genes occurs, while several photosynthesis-related genes were down-regulated [58]. Only minor differences in expression of 15 studied genes were found between the studied cultivars. The similarity in the gene expression of different cultivars tested in the growth chamber and under greenhouse conditions supports the conclusion that the genes identified in response to water withdrawal constitute a general and intrinsic response of common bean to drought and strengthens the relevance of the experimental results to field conditions [58]. In a study on drought tolerant 'Long 22-0579' and drought sensitive 'Naihua' Chinese common bean cultivars, *de novo* assembly of transcriptome data enabled detection of more than nine thousand drought-responsive candidate genes differentially expressed between the drought and control treatments or between both cultivars exposed to drought [56]. Detected genes include those associated with drought-related metabolic processes (cell metabolism, cell wall and carbohydrate biosynthesis), osmoprotectants (proline), transcription factors (MYB, WRKY, DREB, and NAC), plant hormone regulation, signaling, and cell communication. The expression data enabled further characterization of drought responsive NAC transcription factors [64]. In the same two cultivars, 49 novel and 120 known miRNA were detected, 24 of them showing either increased or decreased expression during drought, and only four sharing the same expression pattern between the cultivars [58]. Among the target genes were genes encoding transcription factors, protein kinases and nuclear transcription factors.

Examples of studies aimed at identifying differences in gene expression of particular groups of genes are those focused on aquaporins (AQPs). These are membrane proteins controlling transcellular water movement from the roots and throughout the plant to assimilating tissues. For this reason, they are involved in controlling the ability of plants to regulate their water supply and transport which is closely related to their ability to tolerate or withstand drought. In higher plants, AQPs form a large and diverse protein family with 35 homologs in Arabidopsis (*Arabidopsis thaliana*) and up to 71 homologs in cotton (*Gossypium hirsutum* L.) [65, 66]. In common bean AQPs with the highest mean expression during drought, as well as under normal conditions, were identified [67]. Expression of their genes has been investigated in genotypes with different responses to drought [47, 59, 68]. Increased expression of *PvTIP2;3* was reported in drought-stressed roots of the tolerant genotype [68] and up-regulation of *PvPIP2;5* in leaves of bean exposed to drought was correlated with a reduction in the transpiration rate [69]. In our recent study, physiological measurements indicate greater prevention of water loss in more drought tolerant cultivars, which may be associated with rapid and adequate down-regulation of AQPs in the plasma membrane and tonoplast [47].

4.2 Proteome analysis of drought response

Although studies of drought stress at the gene expression level provide many important data and indications, changes in the transcriptome do not necessarily mean that they will be translated into the proteome level. Studies using proteomic methods are therefore essential for revealing, not only the role of proteins in complex mechanisms of drought response in common bean, but also for pointing out possible molecular markers of drought tolerance. These methods not only enable identification of proteins with abundance changed in response to environmental stress, but also the detection of protein complexes and protein localization, as well as of post-translational protein modifications related to a specific stress factor [70–72] as reported later in this chapter. As underlined above, this approach has experienced rapid development by the recent publishing of full genome sequences of many plants, among them common bean [50].

One of the first studies of drought induced changes in common bean on the proteome was our research focusing on leaves and stems of two cultivars differing in their response to drought [15–17]. In 'Tiber,' we identified 58 proteins whose abundance changed significantly and in 'Starozagorski čern' 64 [15]. Most of the identified proteins were classified into functional categories that include energy metabolism, photosynthesis, ATP interconversion, protein synthesis and proteolysis, stress and defense-related proteins. Significant changes in abundance were observed in large proportion of proteins associated with photosynthesis, such as Rubisco, carbonic anhydrase, oxygen evolving enhancer proteins and chlorophyll a/b binding proteins. While Rubisco small subunit showed lower abundances in drought in both cultivars, carbonic anhydrase was reduced in 'Starozagorski čern,' and in 'Tiber' we detected both increased and reduced abundance. Abundance of chlorophyll a/b binding proteins increased in 'Tiber' and was reduced in 'Starozagorski čern.' The most outstanding contrasting abundance between the two cultivars was the oxygen evolving enhancer proteins, OEE1 and OEE2. Significant changes in abundance were observed in case of a few of the proteins involved in response to stress (e.g., superoxide dismutase, ascorbate peroxidase, and dehydrin) and in case of proteins associated with proteolysis and protein folding (e.g., cysteine proteinase CP2, precursors of cysteine proteinase, proteasome subunit beta type, peptidyl-prolyl cis-trans isomerase, and 20 kDa chaperonin). For peptidyl-prolyl cis-trans isomerase we detected higher abundance in "Tiber' and reduced abundance in 'Starozagorski čern' under drought conditions. From the category ATP interconversion, in both cultivars nucleoside diphosphate kinase (NDPK) significantly increased under drought whereas ATP synthase decreased in abundance. Interactions between identified proteins were demonstrated by bioinformatics analysis, enabling a more complete insight into biological pathways and molecular functions affected by drought stress.

The further study on stem [16] carried out on cultivar 'Tiber' showed changed abundances under drought of proteins that can be classified in the same categories as leaf proteins. The proteins with increased abundance indicate the importance of maintaining protein homeostasis to mitigate this stress. There was increased abundance of proteins involved in protein synthesis, proteolysis and protein folding. Among them, the protein with the greatest abundance was 70 kDa heat shock protein that chaperones the correct folding of proteins [16].

4.3 Postranslational modifications in the response to drought

The complexity of the response of plants to drought is further emphasized by reports indicating that posttranslational modifications (PTMs) of proteins also play an important role. These include covalent modifications of a number of cell proteins that follow protein biosynthesis and are usually catalyzed by enzymes. There are different types of PTMs, among them glycosylation, that are recognized as being very important in plants and in their response to stress [73, 74]. This type of PTM affects protein stability, interaction with other proteins, protein trafficking and, as a consequence, protein activity.

Only a small number of the proteomic studies that have been carried out address specifically protein glycosylation and changes in abundance of glycoproteins in crops under abiotic stress [17]. One of them is our study of glycosylated proteins in leaves of common bean stressed by drought [17]. 'Tiber,' previously identified as relatively tolerant to drought [46, 47], was investigated. Thirty-five glycoproteins with changed abundance were detected. Their structures showed high mannose, complex and hybrid types of N-glycans, most of them being associated with the cell wall (many cell wall-degrading enzymes, such as, β-glucosidase, α-arabinofuranosidase and β-xylosidase, were more abundant under drought), with the stress response (such as, ascorbate oxidase, purple acid phosphatase and reticulin oxidase-like protein that were also more abundant) and with proteolysis and protein folding (such as, the precursor of subtilisin-like serine protease, nicastrin, the precursor of cysteine protease and protein disulfide isomerase that were less abundant) [17].

It follows from the studies reported above that proteome analysis of common bean under drought has revealed the participation of proteins involved in proteolysis [15, 17]. Many proteins, after their synthesis, need to be activated by highly regulated proteolytic cleavage of specific peptide bonds that removes parts of their peptide chains. Activation of regulated proteolysis and simultaneous inhibition of uncontrolled proteolysis are vital for cell survival under dehydration stress. All beneficial changes in metabolism under drought require the active involvement of controlled proteolysis that regulates the turnover rates of specific enzymes and/or proteins involved in cell signaling, and ensures degradation of oxidatively damaged, improperly folded and irreversibly denatured proteins [75, 76]. On the other hand non-specific, uncontrolled proteolysis can be damaging to cells, leading to random breakdown of the majority of cell proteins. Such protein degradation, provoked by drought, results mainly in the disruption of cell membranes and exhibits many features in common with plant senescence [77].

Proteolysis is catalyzed by proteases whose activity is regulated mainly by specific plant protease inhibitors both detected by transcriptomics and/or proteomics [76]. The latter are important, not only for inhibiting proteases activated on drought, but also for osmoprotection, since many of them are highly hydrophilic. The striking diversity of plant proteases and of their inhibitors in each species [76] coupled with the fact that very few of their natural substrates are known [78], complicates research in this field. In addition, it appears that the changes in abundance of many proteases in plants stressed by drought have not been detected by proteomics, due to their low abundance. A combination of proteome analysis and measurement of activities is therefore needed. The proteases most often reported to be involved in the response to drought are cysteine endopeptidases [79, 80], although research on legume plants has indicated participation of other catalytic types of protease, such as serine and aspartic endopeptidases [81, 82].

The response to drought at the level of leaf proteases has been relatively extensively investigated in common bean. Different types of protease have been studied at levels ranging from gene expression to proteolytic activity. In several cases, cultivars with different sensitivities to water deficit have been investigated and changes in proteolytic activity correlated with cultivar sensitivity [46, 54, 82]. In leaf extracts from Brazilian cultivars several endoproteolytic activities with different pH optima were higher in plants under drought. This effect correlated with the level of sensitivity to drought of cultivars [54, 82]. Our research, carried out on cultivars of European origin differing in sensitivity, showed the involvement in response to this stress of different classes of endopeptidases [46]. Increased activities with pH optima in the acid region were observed in leaf extracts of the more sensitive cultivars and were assigned to cysteine and serine proteases. It should be emphasized that differential analysis of leaf proteomes indicated higher abundances of cysteine proteinase precursors in stressed samples [15]. In addition, we have found that the activities of five aminopeptidases in leaves of common bean changed when plants were subjected to drought, this response depending on the developmental stage of the leaves [83].

We have further isolated and characterized, at the protein and gene levels, a protease from the leaves of a common bean that is influenced by drought [81]. It has been classified as a new plant subtilisin-like serine protease. While its gene expression did not change on water deficit, its proteolytic activity did. Further, in common bean leaves an aspartic protease was characterized whose activity was strongly induced on water deficit [82, 84]. It was shown that proteolytic processing of its precursor form was induced by drought, and this, together with the effect of stress on the level of its transcript, led to the suggestion that water deficit regulates activity at both the transcriptional and PTM levels. This response occurs earlier and is stronger in the cultivar more susceptible to drought.

5. Applications of QTL and molecular markers in breeding for drought tolerance

5.1 QTL mapping

Quantitative trait locus (QTL) mapping is an established approach for detecting loci associated with complex quantitative traits, such as, plant tolerance to drought. In common bean multiple populations derived from crosses of susceptible and tolerant parental genotypes, belonging to either a single gene pool, or both Andean and Mesoamerican gene-pools, have been genotyped and genetic linkage maps constructed [18, 85]. Their precision and resolution have been greatly improved by novel sequencing technology and genetic markers, such as, SNPs. For instance, two inter-gene pool populations of 'BAT93' × 'JaloEEP558' and 'DOR364' × 'G19833' have been genotyped repeatedly using a variety of marker systems, ranging from SSR and amplified fragment length polymorphism (AFLP) to SNP [18, 85]. The efforts have culminated in consensus linkage map generation joining both major inter-gene pool maps as well as serving as a core for integration with Mesoamerican linkage map [18]. These approaches have enabled identification of numerous QTLs, controlling resistance to various viral, bacterial and fungal pathogens as well as multigenic traits such as tolerance to drought, biomass production, yield partitioning, and micronutrient accumulation [86, 87].

Drought response-associated QTLs in common bean have been reported in association with yield, phenology, canopy biomass and biomass partitioning. A Mesoamerican and Andean inter-gene pool genetic map with high marker coverage was utilized to detect phenological and seed weight QTLs associated with drought tolerance [88], while intra-gene pool Mesoamerican mapping population has been utilized to identify drought-associated QTL for phenological and yield-related traits [89] as well as QTL for photosynthate acquisition, accumulation and remobilization traits in drought stress [90].

The translation of reported QTLs into practical use has, however, been limited, due to highly variable common bean germplasm and strong influence of the environmental conditions on the presence of minor QTLs. It would be ideal to perform the validation of the QTL in crop production areas. Establishing controlled and uniform growth conditions for evaluation of a large recombinant inbred line (RIL) population, exceeding hundred genotypes, can however prove difficult and not very practical, so a compromise approach for validation of major QTL could consist of testing a subsample of the most diverse RILs for a selected segregation trait in multiple trials sites [5]. Much of the work in QTL mapping and development of drought-tolerant cultivars has been performed based on the traits of the drought resistance sources of Mesoamerican origin, such as those belonging to the race Durango [5].

5.2 Marker-assisted selection and breeding for drought tolerance

For decades DNA markers have been the most widely used molecular markers in crop improvement, due to their abundance and polymorphisms. Marker-assisted selection enables precise and effective selection of common bean genotypes with specific traits and can greatly facilitate the selection process in breeding [91, 92]. These markers are potentially very useful in trait selection and breeding applications, and have been utilized in our procedures, to offer additional informative value on the common bean genotypes included in the breeding program (not published). The advantage of such an approach is that a broad range of economically important traits can be covered, including disease resistance, abiotic stress tolerance, high yield, earliness, phosphorus uptake, and root morphology. However, the practical utilization of molecular markers is at the beginning, also due to lacking validation across the genotypes of the diverse common bean germplasm.

The marker-assisted selection is especially effective in selection for simple and single gene traits, and has been applied for selection for resistance genes for various common bean diseases of viral, bacterial and fungal origin [93, 94]. Selection for quantitative traits such as quantitative resistance or drought tolerance presents a great challenge as it can involve multiple major and minor QTLs controlling the trait [91]. Improved understanding of the complex drought response mechanisms on the level of physiology and molecular biology has enabled identification of potential molecular markers, which could help us distinguish between drought resistant and susceptible genotypes. Among the recently reported markers associated with the drought response in common bean are AQPs whose expression is discussed in the present chapter under Section 4.1. On the other hand several potentially useful molecular markers associated with drought response traits, such as high yield under drought, have been identified using QTL mapping in a segregating RIL populations [89, 90, 95].

Common strategy of common bean breeding programs for resistance to drought is selection of best yielding genotypes that are cultivated in drought-exposed conditions [37, 41]. In addition to that application of novel breeding approaches not frequently used in common bean breeding has been described [5]. Recurrent selection has been utilized for breeding for drought resistance in genotypes within the same gene-pool, following a process of pre-breeding in which multiple potential parental genotypes with drought resistance traits are created [40]. Another breeding method, advanced backcrossing, could be potentially useful for simultaneous transfer of multiple genes for improving drought resistance traits across genepools [5]. These breeding approaches could greatly benefit with the future developments in the research of plant drought response mechanisms and discovery of associated molecular markers.

6. Conclusion

Drought tolerance is gaining importance in the breeding of common bean for higher yields under the changing environmental conditions. Studying drought tolerance is thus important in order to understand the underlying mechanisms and to identify markers that could help distinguish the more tolerant common bean genotypes. A highly diverse common bean germplasm, adapted to various growth and climatic conditions, constitutes a valuable pool of traits including potential drought tolerance traits. On the other hand, the great complexity of the common bean response to drought on physiological and molecular levels presents a great problem for more effective breeding. The challenge for the future will be to integrate the data obtained by various approaches that include screening of the transcriptome, proteome and metabolome, using advanced bioinformatics and systems biology, identifying molecular markers and QTLs and elucidating the underlying pathways.

Acknowledgements

The contribution of Dr. Marko Maras and Dr. Mateja Zupin to our studies presented is greatly acknowledged. The authors thank Professor Roger H. Pain for critically reviewing the manuscript. This work was supported by the Slovenian Research Agency (P4-0072, L4-7520) and the Ministry of Agriculture, Forestry and Food (L4-7520).

Conflict of interest

The authors declare no conflict of interest.

Author details

Aleš Sedlar[1], Marjetka Kidrič[2], Jelka Šuštar-Vozlič[1], Barbara Pipan[1], Tanja Zadražnik[1] and Vladimir Meglič[1]*

1 Agricultural Institute of Slovenia, Ljubljana, Slovenia

2 Jožef Stefan Institute, Ljubljana, Slovenia

*Address all correspondence to: vladimir.meglic@kis.si

IntechOpen

© 2019 The Author(s). Licensee IntechOpen. This chapter is distributed under the terms of the Creative Commons Attribution License (http://creativecommons.org/licenses/by/3.0), which permits unrestricted use, distribution, and reproduction in any medium, provided the original work is properly cited. (cc) BY

References

[1] Broughton WJ, Hernández G, Blair M, Beebe S, Gepts P, Vanderleyden J. Beans (*Phaseolus* spp.)—Model food legumes. Plant and Soil. 2003;**252**:55-128

[2] de Almeida Costa GE, da Silva Queiroz-Monici K, Pissini Machado Reis SM, de Oliveira AC. Chemical composition, dietary fibre and resistant starch contents of raw and cooked pea, common bean, chickpea and lentil legumes. Food Chemistry. 2006;**94**(3):327-330

[3] Blair MW, Sandoval Medina TA, Caldas GV, Beebe SE, Paez MI. Quantitative trait locus analysis of seed phosphorus and seed phytate content in a recombinant inbred line population of common bean. Crop Science. 2009;**49**:237-246. Available from: https://cgspace.cgiar.org/handle/10568/43100

[4] Blair MW, Wu X, Bhandari D, Astudillo C. Genetic dissection of ICP-detected nutrient accumulation in the whole seed of common bean (*Phaseolus vulgaris* L.). Frontiers in Plant Science. 2016;**7**:219

[5] Beebe S, Rao I, Blair M, Acosta J. Phenotyping common beans for adaptation to drought. Plant Physiology. 2013;**4**:35

[6] Rao IM. Role of physiology in improving crop adaptation to abiotic stresses in the tropics: The case of common bean and tropical forages. In: Pessarakli M, editor. Handbook of Plant and Crop Physiology. 2nd ed. New York, NY, USA: Basel; 2001. pp. 583-613

[7] White JW, Singh SP. Sources and inheritance of earliness in tropically adapted indeterminate common bean. Euphytica. 1991;**55**:15-19

[8] Levitt J. Responses of Plants to Environmental Stresses. New York: Academic Press; 1972. p. 698

[9] Leonardis AMD, Petrarulo M, Vita PD, Mastrangelo AM. Genetic and molecular aspects of plant response to drought in annual crop species. In: Montanaro G, editor. Advances in Selected Plant Physiology Aspects. Rijeka: InTech; 2012. Available from: https://www.intechopen.com/books/advances-in-selected-plant-physiology-aspects/genetic-and-molecular-aspects-of-plant-response-to-drought-stress

[10] Arve L, Torre S, Olsen J, Tanino KK. Stomatal responses to drought stress and air humidity. In: Shanker A, editor. Abiotic Stress in Plants—Mechanisms and Adaptations. Rijeka: InTech; 2011. Available from: https://www.intechopen.com/books/abiotic-stress-in-plants-mechanisms-and-adaptations/stomatal-responses-to-drought-stress-and-air-humidity

[11] Chaves MM, Flexas J, Pinheiro C. Photosynthesis under drought and salt stress: Regulation mechanisms from whole plant to cell. Annals of Botany. 2008;**103**(4):551-560

[12] Ramachandra Reddy A, Chaitanya KV, Vivekanandan M. Drought-induced responses of photosynthesis and antioxidant metabolism in higher plants. Journal of Plant Physiology. 2004;**161**:1189-1202

[13] Vaseva I, Akiscan Y, Demirevska K, Anders I, Feller U. Drought stress tolerance of red and white clover—Comparative analysis of some chaperonins and dehydrins. Scientia Horticulturae. 2011;**130**:653-659

[14] Mir RR, Zaman-Allah M, Sreenivasulu N, Trethowan R, Varshney RK. Integrated genomics, physiology and breeding approaches for improving drought tolerance in crops. Theoretical and Applied Genetics. 2012;**125**:625-645

[15] Zadražnik T, Hollung K, Egge-Jacobsen W, Meglič V, Šuštar-Vozlič J. Differential proteomic analysis of drought stress response in leaves of common bean (*Phaseolus vulgaris* L.). Journal of Proteomics. 2013;**78**:254-272

[16] Zadražnik T, Egge-Jacobsen W, Meglič V, Šuštar-Vozlič J. Proteomic analysis of common bean stem under drought stress using in-gel stable isotope labeling. Journal of Plant Physiology. 2017;**209**:42-50

[17] Zadražnik T, Moen A, Egge-Jacobsen W, Meglič V, Šuštar-Vozlič J. Towards a better understanding of protein changes in common bean under drought: A case study of N-glycoproteins. Plant Physiology and Biochemistry. 2017;**118**:400-412

[18] Galeano CH, Fernandez AC, Franco-Herrera N, Cichy KA, McClean PE, Vanderleyden J, et al. Saturation of an intra-gene pool linkage map: Towards a unified consensus linkage map for fine mapping and synteny analysis in common bean. PLoS ONE. 2011;**6**:e28135

[19] Gepts P. Origin and evolution of common bean: Past events and recent trends. Horticultural Science. 1998;**33**:1124-1130

[20] Bitocchi E, Nanni L, Bellucci E, Rossi M, Giardini A, Zeuli PS, et al. Mesoamerican origin of the common bean (*Phaseolus vulgaris* L.) is revealed by sequence data. Proceedings of the National Academy of Sciences. 2012;**109**:E788-E796

[21] Gepts P, Bliss FA. Phaseolin variability among wild and cultivated common beans (*Phaseolus vulgaris*) from Colombia. Economic Botany. 1986;**40**:469-478

[22] Sinkovič L, Pipan B, Sinkovič E, Meglič V. Morphological seed characterization of common (*Phaseolus vulgaris* L.) and runner (*Phaseolus coccineus* L.) bean germplasm: A Slovenian Gene Bank example. BioMed Research International. 2019;**2019**:Article ID 6376948. Available from: https://www.hindawi.com/journals/bmri/2019/6376948/

[23] Asfaw A, Blair MW, Almekinders C. Genetic diversity and population structure of common bean (*Phaseolus vulgaris* L.) landraces from the East African highlands. Theoretical and Applied Genetics. 2009;**120**:1-12

[24] Blair MW, González LF, Kimani PM, Butare L. Genetic diversity, inter-gene pool introgression and nutritional quality of common beans (*Phaseolus vulgaris* L.) from Central Africa. Theoretical and Applied Genetics. 2010;**121**(2):237-248

[25] Angioi SA, Rau D, Attene G, Nanni L, Bellucci E, Logozzo G, et al. Beans in Europe: Origin and structure of the European landraces of *Phaseolus vulgaris* L. Theoretical and Applied Genetics. 2010;**121**:829-843

[26] Angioi SA, Rau D, Nanni L, Bellucci E, Papa R, Attene G. The genetic make-up of the European landraces of the common bean. Plant Genetic Resources. 2011;**9**:197-201

[27] Maras M, Šuštar-Vozlič J, Kainz W, Meglič V. Genetic diversity and dissemination pathways of common bean in Central Europe. Journal of the American Society for Horticultural Science. 2013;**138**:297-305

[28] Maras M, Pipan B, Šuštar-Vozlič J, Todorović V, Đurić G, Vasić M, et al. Examination of genetic diversity of common bean from the Western Balkans. Journal of the American Society for Horticultural Science. 2015;**140**:308-316

[29] Acosta-Gallegos JA, Kelly JD, Gepts P. Prebreeding in common bean and use of genetic diversity from wild germplasm. Crop Science. 2007;**47**:S44-S59

[30] Šuštar-Vozlič J, Maras M, Javornik B, Meglič V. Genetic diversity and origin of Slovene common bean (*Phaseolus vulgaris* L.) germplasm as revealed by AFLP markers and phaseolin analysis. Journal of the American Society for Horticultural Science. 2006;**131**:242-249

[31] Dwivedi SL, Ceccarelli S, Blair MW, Upadhyaya HD, Are AK, Ortiz R. Landrace germplasm for improving yield and abiotic stress adaptation. Trends in Plant Science. 2016;**21**:31-42

[32] Pipan B, Dolnicar P, Sedlar A, Šuštar Vozlič J, Sinkovič L, Meglič V. Application of molecular data to construct common bean core collection (Central and South Eastern European germplasm example). In: Poster Presentation Abstracts. Poster Presentation Abstracts. Vol. 2018. Rotterdam. p. 7

[33] da Silva EC, de Albuquerque MB, de Azevedo Neto AD, da Silva Junior CD. Drought and its consequences to plants—From individuals to ecosystem. In: Akinci S, editor. Responses of Organisms to Water Stress. Rijeka: Intech; 2013. Available from: https://www.intechopen.com/books/responses-of-organisms-to-water-stress/drought-and-its-consequences-to-plants-from-individual-to-ecosystem

[34] de Oliveira AB, Mendes Alencar NL, Gomes-Filho E. Comparison between the Water and Salt Stress Effects on Plant Growth and Development. Rijeka: Intech; 2013. https://www.intechopen.com/books/responses-of-organisms-to-water-stress/comparison-between-the-water-and-salt-stress-effects-on-plant-growth-and-development

[35] Shaar-Moshe L, Blumwald E, Peleg Z. Unique physiological and transcriptional shifts under combinations of salinity, drought, and heat. Plant Physiology. 2017;**174**:421-434

[36] Isayenkov SV, Maathuis FJM. Plant salinity stress: Many unanswered questions remain. Frontiers in Plant Science. 2019;**10**:80

[37] Polania J, Rao IM, Cajiao C, Rivera M, Raatz B, Beebe S. Physiological traits associated with drought resistance in Andean and Mesoamerican genotypes of common bean (*Phaseolus vulgaris* L.). Euphytica. 2016;**210**:17-29

[38] Costa França M, Pham Thi A, Pimentel C, Pereyra Rossiello R, Zuily-Fodil Y, Laffray D. Differences in growth and water relations among *Phaseolus vulgaris* cultivars in response to induced drought stress. Environmental and Experimental Botany. 2000;**43**:227-237

[39] Lizana C, Wentworth M, Martinez JP, Villegas D, Meneses R, Murchie EH, et al. Differential adaptation of two varieties of common bean to abiotic stress: I. Effects of drought on yield and photosynthesis. Journal of Experimental Botany. 2006;**57**:685-697

[40] Beebe SE, Rao IM, Cajiao C, Grajales M. Selection for drought resistance in common bean also improves yield in phosphorus limited and favorable environments. Crop Science. 2008;**48**:582-592

[41] Rosales MA, Ocampo E, Rodríguez-Valentín R, Olvera-Carrillo Y, Acosta-Gallegos J, Covarrubias AA. Physiological analysis of common bean (*Phaseolus vulgaris* L.) cultivars uncovers characteristics related to terminal drought resistance. Plant Physiology and Biochemistry. 2012;**56**:24-34

[42] Rosales MA, Cuellar-Ortiz SM, de la Paz Arrieta-Montiel M, Acosta-Gallegos J, Covarrubias AA. Physiological traits related to terminal drought resistance in common bean (*Phaseolus vulgaris* L.). Journal of the Science of Food and Agriculture. 2013;**93**:324-331

[43] Rao I, Beebe S, Polania J, Ricaurte J, Cajiao C, Garcia R, et al. Can tepary bean be a model for improvement of drought resistance in common bean? African Crop Science Journal. 2013;**21**:265-281. Available from: https://www.ajol.info/index.php/acsj/article/view/95291

[44] Ruiz-Nieto JE, Aguirre-Mancilla CL, Acosta-Gallegos JA, Raya-Pérez JC, Piedra-Ibarra E, Vázquez-Medrano J, et al. Photosynthesis and chloroplast genes are involved in water-use efficiency in common bean. Plant Physiology and Biochemistry. 2015;**86**:166-173

[45] Chaves MM, Pereira JS, Maroco J, Rodrigues ML, Ricardo CPP, Osório ML, et al. How plants cope with water stress in the field. Photosynthesis and growth. Annals of Botany. 2002;**89** (Spec No):907-916

[46] Hieng B, Ugrinović K, Šuštar-Vozlič J, Kidrič M. Different classes of proteases are involved in the response to drought of *Phaseolus vulgaris* L. cultivars differing in sensitivity. Journal of Plant Physiology. 2004;**161**:519-530

[47] Zupin M, Sedlar A, Kidrič M, Meglič V. Drought-induced expression of aquaporin genes in leaves of two common bean cultivars differing in tolerance to drought stress. Journal of Plant Research. 2017;**130**:735-745

[48] Young ND, Debellé F, Oldroyd GED, Geurts R, Cannon SB, Udvardi MK, et al. The *Medicago* genome provides insight into the evolution of rhizobial symbioses. Nature. 2011;**480**:520-524

[49] Sato S, Nakamura Y, Kaneko T, Asamizu E, Kato T, Nakao M, et al. Genome structure of the legume, *Lotus japonicus*. DNA Research. 2008;**15**:227-239

[50] Schmutz J, McClean PE, Mamidi S, Wu GA, Cannon SB, Grimwood J, et al. A reference genome for common bean and genome-wide analysis of dual domestications. Nature Genetics. 2014;**46**:707-713

[51] Schmutz J, Cannon SB, Schlueter J, Ma J, Mitros T, Nelson W, et al. Genome sequence of the palaeopolyploid soybean. Nature. 2010;**463**:178-183

[52] Jain M, Misra G, Patel RK, Priya P, Jhanwar S, Khan AW, et al. A draft genome sequence of the pulse crop chickpea (*Cicer arietinum* L.). Plant Journal for Cell and Molecular Biology. 2013;**74**:715-729

[53] Bertioli DJ, Cannon SB, Froenicke L, Huang G, Farmer AD, Cannon EKS, et al. The genome sequences of *Arachis duranensis* and *Arachis ipaensis*, the diploid ancestors of cultivated peanut. Nature Genetics. 2016;**48**:438-446

[54] Roy-Macauley H, Zuily-Fodil Y, Kidric M, Thi ATP, de Silva JV. Effect of drought stress on proteolytic activities in *Phaseolus* and *Vigna* leaves from sensitive and resistant plants. Physiologia Plantarum. 1992;**85**:90-96

[55] Torres GAM, Pflieger S, Corre-Menguy F, Mazubert C, Hartmann C, Lelandais-Brière C. Identification of novel drought-related mRNAs in common bean roots by differential display RT-PCR. Plant Science. 2006;**171**:300-307

[56] Wu J, Wang L, Li L, Wang S. *De novo* assembly of the common bean transcriptome using short reads for the discovery of drought-responsive genes. Chen Z-H, editor. PLoS One. 2014;**9**:e109262

[57] Sengupta D, Kannan M, Reddy AR. A root proteomics-based insight reveals dynamic regulation of root proteins under progressive drought stress and recovery in *Vigna radiata* (L.) Wilczek. Planta. 2011;**233**:1111-1127

[58] Kavar T, Maras M, Kidrič M, Šuštar-Vozlič J, Meglič V. Identification of genes

involved in the response of leaves of *Phaseolus vulgaris*. Molecular Breeding. 2007;**11**(21):159-172

[59] Recchia GH, Caldas DGG, Beraldo ALA, da Silva MJ, Tsai SM. Transcriptional analysis of drought-induced genes in the roots of a tolerant genotype of the common bean (*Phaseolus vulgaris* L.). International Journal of Molecular Sciences. 2013;**14**:7155-7179

[60] Arenas-Huertero C, Pérez B, Rabanal F, Blanco-Melo D, De la Rosa C, Estrada-Navarrete G, et al. Conserved and novel miRNAs in the legume *Phaseolus vulgaris* in response to stress. Plant Molecular Biology. 2009;**70**:385-401

[61] Sosa-Valencia G, Palomar M, Covarrubias AA, Reyes JL. The legume miR1514a modulates a NAC transcription factor transcript to trigger phasiRNA formation in response to drought. Journal of Experimental Botany. 2017;**68**(8):2013-2026

[62] Wu J, Wang L, Wang S. MicroRNAs associated with drought response in the pulse crop common bean (*Phaseolus vulgaris* L.). Gene. 2017;**628**:78-86

[63] Torres GAM, Lelandais-Brière C, Besin E, Jubier M-F, Roche O, Mazubert C, et al. Characterization of the expression of *Phaseolus vulgaris* OCT1, a dehydration-regulated gene that encodes a new type of phloem transporter. Plant Molecular Biology. 2003;**51**:341-349

[64] Wu J, Wang L, Wang S. Comprehensive analysis and discovery of drought-related NAC transcription factors in common bean. BMC Plant Biology. 2016;**16**:193. Available from: https://www.ncbi.nlm.nih.gov/pmc/articles/PMC5013670/

[65] Johanson U, Karlsson M, Johansson I, Gustavsson S, Sjövall S, Fraysse L, et al. The complete set of genes encoding major intrinsic proteins in *Arabidopsis* provides a framework for a new nomenclature for major intrinsic proteins in plants. Plant Physiology. 2001;**126**:1358-1369

[66] Park W, Scheffler BE, Bauer PJ, Campbell BT. Identification of the family of aquaporin genes and their expression in upland cotton (*Gossypium hirsutum* L.). BMC Plant Biology. 2010;**10**:142

[67] Ariani A, Gepts P. Genome-wide identification and characterization of aquaporin gene family in common bean (*Phaseolus vulgaris* L.). Molecular Genetics and Genomics. 2015;**290**:1771-1785

[68] Montalvo-Hernández L, Piedra-Ibarra E, Gómez-Silva L, Lira-Carmona R, Acosta-Gallegos JA, Vazquez-Medrano J, et al. Differential accumulation of mRNAs in drought-tolerant and susceptible common bean cultivars in response to water deficit. The New Phytologist. 2008;**177**:102-113

[69] Aroca R, Porcel R, Ruiz-Lozano JM. Regulation of root water uptake under abiotic stress conditions. Journal of Experimental Botany. 2012;**63**:43-57

[70] Abreu IA, Farinha AP, Negrão S, Gonçalves N, Fonseca C, Rodrigues M, et al. Coping with abiotic stress: Proteome changes for crop improvement. Journal of Proteomics. 2013;**93**:145-168

[71] Kosová K, Vítámvás P, Prášil IT, Renaut J. Plant proteome changes under abiotic stress-contribution of proteomics studies to understanding plant stress response. Journal of Proteomics. 2011;**74**:1301-1322

[72] Kosová K, Vítámvás P, Urban MO, Prášil IT, Renaut J. Plant abiotic stress proteomics: The major factors determining alterations in cellular proteome. Frontiers in Plant Science. 2018;**9**:122

[73] Fitchette A-C, Dinh OT, Faye L, Bardor M. Plant proteomics and glycosylation. Methods in Molecular Biology (Clifton, N.J.). 2007;**355**:317-342

[74] Lannoo N, Van Damme EJM. Lectin domains at the frontiers of plant defense. Frontiers in Plant Science. 2014;**5**:397

[75] Vaseva I, Sabotič J, Šuštar-Vozlič J, Meglič V, Kidrič M, Demirevska K, et al. The response of plants to drought stress: The role of dehydrins, chaperones, proteases and protease inhibitors in maintaining cellular protein function. In: Neves DF, Sanz JD, editors. Droughts: New Research. New York: Nova Science Publishers; 2012

[76] Kidrič M, Kos J, Sabotič J. Proteases and their endogenous inhibitors in the plant response to abiotic stress. Botanica Serbica. 2014;**38**(1):139-158

[77] Khanna-Chopra R. Leaf senescence and abiotic stresses share reactive oxygen species-mediated chloroplast degradation. Protoplasma. 2012;**249**(3):469-481

[78] Tsiatsiani L, Gevaert K, Van Breusegem F. Natural substrates of plant proteases: How can protease degradomics extend our knowledge? Physiologia Plantarum. 2012;**145**(1):28-40

[79] Ingram J, Bartels D. The molecular basis of dehydration tolerance in plants. Annual Review of Plant Physiology and Plant Molecular Biology. 1996;**47**:377-403

[80] Simova-Stoilova L, Vaseva I, Grigorova B, Demirevska K, Feller U. Proteolytic activity and cysteine protease expression in wheat leaves under severe soil drought and recovery. Plant Physiology and Biochemistry. 2010;**48**:200-206

[81] Budič M, Sabotič J, Meglič V, Kos J, Kidrič M. Characterization of two novel subtilases from common bean (*Phaseolus vulgaris* L.) and their responses to drought. Plant Physiology and Biochemistry. 2013;**62**:79-87

[82] Cruz de Carvalho MH, d'Arcy-Lameta A, Roy-Macauley H, Gareil M, El Maarouf H, Pham-Thi AT, et al. Aspartic protease in leaves of common bean (*Phaseolus vulgaris* L.) and cowpea (*Vigna unguiculata* L. Walp): Enzymatic activity, gene expression and relation to drought susceptibility. FEBS Letters. 2001;**492**(3):242-246

[83] Budič M, Cigić B, Sostaric M, Sabotič J, Meglič V, Kos J, et al. The response of aminopeptidases of *Phaseolus vulgaris* to drought depends on the developmental stage of the leaves. Plant Physiology and Biochemistry. 2016;**109**:326-336

[84] Contour-Ansel D, Torres-Franklin ML, Zuily-Fodil Y, de Carvalho MHC. An aspartic acid protease from common bean is expressed "on call" during water stress and early recovery. Journal of Plant Physiology. 2010;**167**:1606-1612

[85] Hanai LR, Santini L, Camargo LEA, Fungaro MHP, Gepts P, Tsai SM, et al. Extension of the core map of common bean with EST-SSR, RGA, AFLP, and putative functional markers. Molecular Breeding. 2010;**25**:25-45

[86] Collins NC, Tardieu F, Tuberosa R. Quantitative trait loci and crop performance under abiotic stress: Where do we stand? Plant Physiology. 2008;**147**:469-486

[87] Guzman-Maldonado SH, Martinez O, Acosta-Gallegos JA, Guevara-Lara F, Paredes-Lopez O. Putative quantitative trait loci for physical and chemical components of common bean. Crop Science. 2003;**43**:1029-1035

[88] Mukeshimana G, Lasley AL, Loescher WH, Kelly JD. Identification of

shoot traits related to drought tolerance in common bean seedlings. Journal of the American Society for Horticultural Science. 2014;**139**:299-309

[89] Blair MW, Galeano CH, Tovar E, Muñoz Torres MC, Castrillón AV, Beebe SE, et al. Development of a Mesoamerican intra-genepool genetic map for quantitative trait loci detection in a drought tolerant × susceptible common bean (*Phaseolus vulgaris* L.) cross. Molecular Breeding. 2012;**29**:71-88

[90] Asfaw A, Blair MW, Struik PC. Multienvironment quantitative trait loci analysis for photosynthate acquisition, accumulation, and remobilization traits in common bean under drought stress. G3 Bethesda Md. 2012;**2**:579-595

[91] Assefa T, Beebe SE, Rao IM, Cuasquer JB, Duque MC, Rivera M, et al. Pod harvest index as a selection criterion to improve drought resistance in white pea bean. Field Crops Research. 2013;**148**:24-33

[92] Varshney RK, Mohan SM, Gaur PM, Gangarao NVPR, Pandey MK, Bohra A, et al. Achievements and prospects of genomics-assisted breeding in three legume crops of the semi-arid tropics. Biotechnology Advances. 2013;**31**:1120-1134

[93] Blair MW, Fregene MA, Beebe SE, Ceballos H. Marker-assisted selection in common beans and cassava. In: Organización de las Naciones Unidas para la Agricultura y la Alimentación (FAO); 2007. Available from: https://cgspace.cgiar.org/handle/10568/55541

[94] Tryphone GM, Chilagane LA, Protas D, Kusolwa PM, Nchimbi-Msolla S. Marker assisted selection for common bean diseases improvements in Tanzania: Prospects and future needs. In: Andersen SB, editor. Plant Breeding from Laboratories to Fields. Rijeka: Intech; 2013. Available from: https://www.intechopen.com/books/plant-breeding-from-laboratories-to-fields/marker-assisted-selection-for-common-bean-diseases-improvements-in-tanzania-prospects-and-future-nee

[95] Briñez B, Perseguini JMKC, Rosa JS, Bassi D, Gonçalves JGR, Almeida C, et al. Mapping QTLs for drought tolerance in a SEA 5 x AND 277 common bean cross with SSRs and SNP markers. Genetics and Molecular Biology. 2017;**40**:813-823

Chapter 8

Understanding the Drought Phenomenon in the Iberian Peninsula

Matilde García-Valdecasas Ojeda, Emilio Romero Jiménez, Sonia R. Gámiz-Fortis, Yolanda Castro-Díez and María Jesús Esteban Parra

Abstract

The analysis and understanding of drought phenomenon are essential for the management of hydrological resources. Drought indices are commonly used to predict these extreme events, being their suitability partly due to the use of climate fields at an adequate spatiotemporal resolution. This work aims to examine spatiotemporal patterns of drought over the Iberian Peninsula (IP), which is a region especially vulnerable to drought phenomenon. For this, climate data from a simulation completed with the Weather Research and Forecasting (WRF) model have been used. The spatiotemporal patterns of drought over the period 1980–2014 were examined using the Standardized Precipitation Evapotranspiration Index (SPEI) at the 3- and the 12-month time scales, and they were compared with other drought-related variables such as the surface evapotranspiration (SFCEVP), soil moisture (SM), and runoff. The results evidence that WRF is a valuable tool for characterizing droughts over the IP, providing large amounts of climate data at an adequate spatial resolution. Drought events seem to be more severe in regard to their duration over southern IP. Moreover, a good agreement between the SPEI at 3-month time scale with the SM and the SFCEVP is found. Additionally, the annual runoff evolves similarly to the SPEI at 12-month time scale.

Keywords: Weather Research and Forecasting model, regional climate models, drought indices, Standardized Precipitation Evapotranspiration Index, Iberian Peninsula

1. Introduction

Drought phenomenon has been considered as one of the major natural hazards causing numerous economic, social, and environmental losses [1]. In the context of ongoing global warming, an increase in frequency and intensity of this kind of extreme event is expected [2–4], and therefore, its accurate characterization is of high relevance for both mitigation and adaptation. In broad terms, drought is usually characterized as the scarcity of precipitation over a prolonged time period, but it is difficult to define and identify due to the great number of variables involved as well as the variety of sectors affected. Moreover, the characterization of the spatiotemporal patterns of drought is very complex since it is very variable in space and time, this being particularly true in transitional zones such as the Iberian Peninsula (IP) [5].

In recent years, drought indices have been commonly used to identify, analyze, and monitor the occurrence of droughts. As indicated in [6], drought indices are variables based on climate information (e.g., precipitation, evapotranspiration, soil moisture, or runoff). They are used to analyze the effects of drought, allowing the definition of different drought characteristics (i.e., the duration and severity of droughts as well as the spatial extent). However, their accuracy strongly depends on long-term consistent climate data, and unfortunately, spatially and temporally regular climate observations are rare. In this context, the regional climate models (RCMs) are valuable tools providing climate information at an adequate spatiotemporal resolution to characterize regional drought patterns. In fact, drought phenomena are spatially complex, so detailed spatial scales are required in the study of droughts [7]. Additionally, Abatzoglou et al. [8] investigated the sensibility of drought indicators to the spatial resolution and found that the indices computed with the highest resolution explained over 10% more variability than those from coarser datasets.

Among different drought indices developed in recent years, the Standardized Precipitation Evapotranspiration Index (SPEI) [9] was proposed as an alternative to the Standardized Precipitation Index (SPI) [10]. Contrariwise to the SPI, the SPEI takes into account the effect of temperature for detecting droughts, and therefore, it seems to be more accurate in the context of global warming [11]. In fact, the increased global temperature trend is expected to increase the atmospheric evaporative demand, so regions, where the precipitation is normal in a given period (or even higher than normal), could be considered to suffer droughts. For instance, the role of the temperature is clear for the event occurred during summer 2003, which had devastating effects in central Europe mainly because of the anomalous temperatures in this period [9]. Furthermore, taking into account temperature data, this drought indicator has shown better performance than others based solely on precipitation (e.g., the SPI) for detecting droughts during summer, when the related impacts may become stronger [11].

This work investigates spatiotemporal patterns of drought through the SPEI to understand the drought temporal behavior over the IP, identifying those periods especially accused. This is of major relevance because the IP is characterized by highly variable and scarce precipitation leading to recurrent drought occurrences [12]. For this purpose, a 35-year climate simulation has been completed using the Weather Research and Forecasting (WRF) model [13] with the purpose of obtaining high-resolution climate data.

The chapter is structured as follows: Section 2 is devoted to detail the WRF model configuration as well as the statistical method applied to characterize the spatiotemporal patterns of droughts over the study region. Section 3 presents the main results regarding the temporal evolution of drought for the 35-year period (1980–2014), the identification of drought characteristics as well as the analysis of the relationships of the SPEI with other drought-related variables, such as the surface (actual) evapotranspiration, the soil moisture content, and the runoff. Finally, Section 4 summarizes the main conclusions obtained in this study.

2. Methodology

2.1 Climate input data

Hydroclimate variables at high spatiotemporal resolution were obtained by using the WRF-ARW model version 3.6.1. This RCM was run to simulate current climate characteristics over the IP and Balearic Islands (**Figure 1a**). The WRF model was driven by the ECMWF ERA-Interim reanalysis data [14]. **Table 1** summarizes the main model setup used in this study. This configuration has been successfully used to represent the spatiotemporal patterns of droughts in the study region [15]. In [15], the authors analyzed the added value of using downscaled climate data to detect drought through the comparison with observational data and found that WRF provided a benefit with respect to its driving data in this regard.

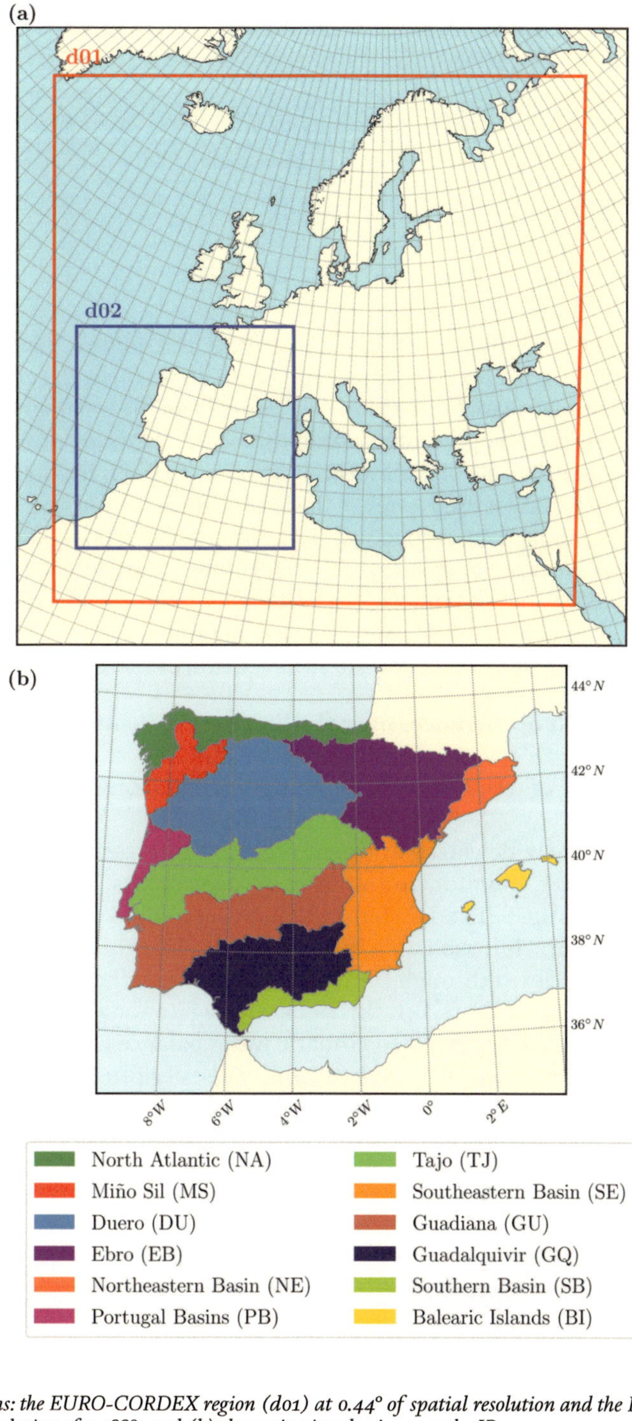

Figure 1.
(a) WRF domains: the EURO-CORDEX region (d01) at 0.44° of spatial resolution and the IP region (d02) with a spatial resolution of 0.088°, and (b) the main river basins over the IP.

Parameters	Description
Spatial configuration	Two "one-way" nested domains: (d01) EURO-CORDEX region at 0.44° (~50 km) of spatial resolution, and (d02) the IP at a spatial resolution of 0.088° (~10 km)
Temporal period	Period 1980–2014 with 11 months of spin-up
Vertical layer	41 vertical levels set the top of the atmosphere at 10 hPa
Nudging and sponge zone	Spectral nudging for waves above 600 km, only over the coarser domain and above the planetary boundary (PBL) Sponge zone: the 10 outer grid points of each domain
Physics schemes	Microphysics: WRF single-moment-3-class [16] Convection: Betts-Miller-Janjic [17, 18] PBL: convective asymmetric model version 2 [19] Land surface model: Noah LSM [20] Radiation: community atmosphere model 3.0 [21]

See [15] for more details.

Table 1.
The main model configuration.

In this study, the 3-hourly outputs in their native grid resolution have been used as climatic input data to analyze drought occurrences. In this way, the outputs of the WRF model at 10 km of spatial resolution (i.e., those from the inner domain d02, **Figure 1a**) were temporally aggregated at monthly scale, obtaining thus gridded long-term monthly climate data over land for the entire study region, for the period 1980–2014. The monthly variables used here are the averaged maximum and minimum temperature (Tmax and Tmin, respectively), the accumulated precipitation (pr), the accumulated surface evapotranspiration (SFCEVP), the mean soil moisture contained in the upper 1 m (SM), and the runoff. The WRF model, through the coupled Noah LSM scheme, contemplates two different runoff components, the surface and subsurface runoff. Here, the runoff was understood as the sum of these two components.

2.2 Defining drought occurrences over the Iberian Peninsula

2.2.1 The Standardized Precipitation Evapotranspiration Index (SPEI)

The SPEI has been widely used in recent years to characterize droughts, showing a good ability to detect, monitor, and analyze drought events [22–24]. This drought indicator is based on the SPI and differs in that it uses a simple "climatic water balance" instead of precipitation data in determining droughts. The temperature effect is indeed considered through the difference between the precipitation and the reference evapotranspiration (ET_0), this being aggregated at a range of time scales (1–48 months). Then, the aggregated climatic water balance is fitted to a statistical distribution, and subsequently, its probability density function is transformed into a random normal variable Z with mean of 0 and variance of 1. This latter assumption enables the comparison across regions by determining the probability of occurrence of a given drought event (**Table 2**). Thus, Z is the corresponding SPEI and indicates the number of standard deviations from the climatological mean. Therefore, and similar to the SPI, the SPEI is statistically robust, simple to compute and easily interpretable. Even more, the option to compute it at different time aggregations makes this index suitable for assessing different drought types (i.e., meteorological, agricultural, or hydrological droughts) [9].

In this study, the SPEI was computed at two different time aggregations: the 3- and 12-month time scales (hereinafter SPEI-03 and SPEI-12, respectively) using the SPEI R-package [25]. These time scales were chosen with the purpose of characterizing

SPEI value	Category	Probability (%)
SPEI ≥2	Extremely wet	2.3
2 ≥ SPEI >1.5	Severely wet	4.4
1.5 ≥ SPEI >1	Moderately wet	9.2
1 ≥ SPEI >0.5	Mildly wet	15.0
0.5 ≥ SPEI > −0.5	Nearly normal	38.2
−0.5 ≥ SPEI > −1.0	Mild drought	15.0
−1 ≥ SPEI > −1.5	Moderate drought	9.2
−1.5 ≥ SPEI > −2	Severe drought	4.4
SPEI < −2	Extreme drought	2.3

Table 2.
Drought categories and their probability of occurrence.

agricultural and hydrological droughts [6], respectively. Temporal series of the climatic water balance were obtained for the entire IP using the pr and the Tmax and Tmin from WRF at monthly scale. To approximate the ET_0, the Hargreaves equation [26] was used. Studies such as [27] evidenced that the Hargreaves method is able to approximate the ET_0 showing similar results to the Penman-Monteith equation [28] with the advantage that only temperature data are required to calculate it. The climatic water balance, here, is assumed to follow a log-logistic distribution [27]. This distribution has been used in many studies (e.g., [10, 27]) to fit the SPEI, mostly because it is a three-parameter distribution (i.e., negative values are permitted), which has been proved to better estimate the climatic water balance [27].

Then, the temporal series of SPEI at both 3- and 12-month time scales for each grid point were spatially averaged obtaining the SPEI evolution for each main river basin over the IP. Twelve main river basins were considered, as the result of aggregating smaller watersheds. They are North Atlantic (NA; composed by the Galician coast, the Western Cantabrian, and the Eastern Cantabrian watersheds), Miño-Sil (MS; the Miño-Sil, the Cávado, the Ave, and the Leça watersheds), Duero (DU), Ebro (EB), Northeastern Basins (NE), Portugal Basins (PB; the Vouga, Mondego, Lis, and Ribeiras do Oeste), Tajo (TJ), Southeastern Basins (SE), Guadiana (GU; Guadiana, Sado, Mira, and Ribeiras do Algarve), Guadalquivir (GQ; Guadalquivir, Tinto, Odiel, Piedras, Guadalete, and Barbate), Southern Basins (SB), and the Balearic Islands (BI) (**Figure 1b**).

In order to explore the severity of droughts, different drought characteristics have also been analyzed. For this, the established condition is that a dry event is occurring when at least two consecutive months are under the defined drought conditions. Therefore, a dry event was considered to begin when the SPEI falls below zero, and it ends when recovering positive values. Moreover, such a dry period is defined as drought if at least 1 month within the period reaches mild drought conditions (i.e., SPEI below −0.5 [29]). Then, the duration of droughts is understood as the number of months in each drought event. The intensity is the averaged value of the SPEI within the period expressed in absolute value. Additionally, the severity of the event (or minimum value reached) was also explored, which is expressed in terms of absolute value of the SPEI.

2.2.2 Temporal evolution of drought-related variables

Time series for each river basin from other drought-related variables were also examined. In this regard, the SM and SFCEVP were examined because they directly affect the agricultural droughts. In fact, soil water availability is the major driver of the plant transpiration. To do this, the monthly time series of SM and SFCEVP were used to compute the standardized anomalies of such variables. These anomalies were computed with two purposes: (1) to remove seasonality and (2) to make the temporal evolutions comparable across regions. The SFCEVP was previously accumulated using 3-month time slices to also compare it with the SPEI-03, which is the time scale here used to characterize agricultural droughts.

On the other hand, to further investigate hydrological droughts, the temporal series of runoff, previously aggregated at annual scale, were used to compute the standardized anomalies series. This variable allows us to incorporate hydrological processes in drought characterization [6]. Hydrological droughts are developed slowly and persist longer than other forms of drought [30], so relationships between temporal series of runoff and the SPEI-12 must occur.

The relationship between such drought-related variables with the SPEI was investigated by computing temporal correlation coefficients for all river basins. Additionally, a t-test at the 95% confidence level was used to determine the significance of the correlation coefficients, previously considering the effect of the serial correlation by following the methodology proposed by Bretherton et al. [31].

3. Spatiotemporal patterns of droughts

3.1 Temporal evolution of the SPEI

This study begins with the analysis of the temporal evolution of the SPEI-03 and the SPEI-12 for all river basins over the entire study period (**Figure 2**). In the interest of clarity, SPEI evolution is represented according to the different drought categories (see **Table 2**). Concerning the SPEI-03 evolution (**Figure 2a**), the results showed a high temporal variability, which is a characteristic of meteorological droughts. That is, the SPEI changes frequently between drought and wet conditions. Furthermore, these results evidence the complexity of drought phenomena and therefore its characterization since very different conditions may be found in the different watersheds in a given moment (i.e., while certain river basins are affected by drought conditions, others presented normal or wet conditions). Several drought events were recorded over the study period. For instance, moderate to extreme drought conditions appeared during the years 1985, 1994–1995, 2005, and 2012 across a large part of the IP. Contrariwise, the years 1984, 1996–1998, 2003–2004 2010, and 2013 showed wet conditions over a large part of the IP.

Looking at the longest time scale (**Figure 2b**), the results presented a less variable behavior. In fact, for 12 months, the SPEI is less sensitive to variations from a given month, and hence, the changes between wet and dry conditions are less frequent and these are also longer. Moreover, the results display that, in general, dry and wet conditions are allocated in the same periods than those from the 3-month time scale, with the most severe droughts happening during 1995, 1999, 2005, and 2012. For these years, many watersheds were affected by severe drought conditions (i.e., SPEI below −2). Conversely, wet periods also appeared over many river basins during the years 1984, 1997, 2010, and 2013. Note the extreme wet conditions occurring during 2001 in the PB and the MS river basins.

3.2 Climatological characteristics of droughts

In order to further explore drought phenomenon, the duration distributions of drought events were explored for all river basins. **Figure 3** displays the box

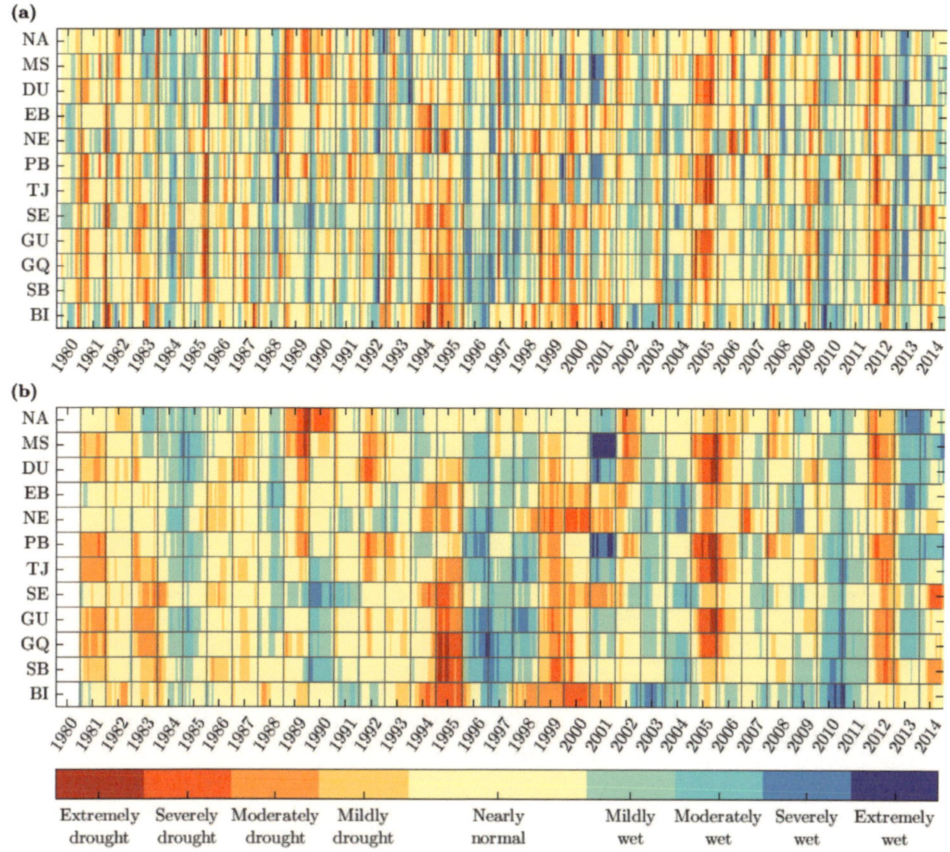

Figure 2.
Temporal evolution of the SPEI in all river basins between 1980 and 2014 at (a) 3- and (b) 12-month time scales. The color code is established according to the drought categories. Nomenclature: North Atlantic (NA), Miño-Sil (MS), Duero (DU), Ebro (EB), Northeastern Basins (NE), Portugal Basins (PB), Tajo (TJ), Southeastern Basins (SE), Guadiana (GU), Guadalquivir (GQ), Southern Basins (SB), and the Balearic Islands (BI).

plots of the duration of droughts for both the SPEI-03 and the SPEI-12 and for all river basins. This representation shows the scores of duration (in months) as well as their distributional characteristics. At 3 months (**Figure 3a**), the median was between 3 and 4 months, and the mean was around 4 months. Therefore, the mean was slightly higher than the median in all river basins, showing that the distributions are skewed to the right. The 97.5th percentile (limits of the upper whiskers) ranges from about 7.5 to 14.5 months, with the shortest and the longest durations occurring for the NE and the GQ river basins, respectively. On the other hand, the spread of the distribution is a signal of high variability in drought events, and it is marked by the size of the box (i.e., the interquartile range). In this regard, southern river basins (i.e., the TJ, SE, GU, GQ, and SB river basins) showed larger interquartile range than the northern ones (i.e., the NA, MS, DU, NE, and PB river basins).

For the SPEI-12 (**Figure 3b**), longer durations appeared for the period 1980–2014 in general, showing also more spread in their distributions. The median was between 5 and 14 months, reaching the maximum values again over the southern river basins. The mean duration of droughts was around 12 months, indicating that the mean was higher than the median in many cases. The 97.5th percentile was between 12 and 53 months with the highest one over the BI river basin. The major variability in terms of interquartile range seems to appear again over the BI river

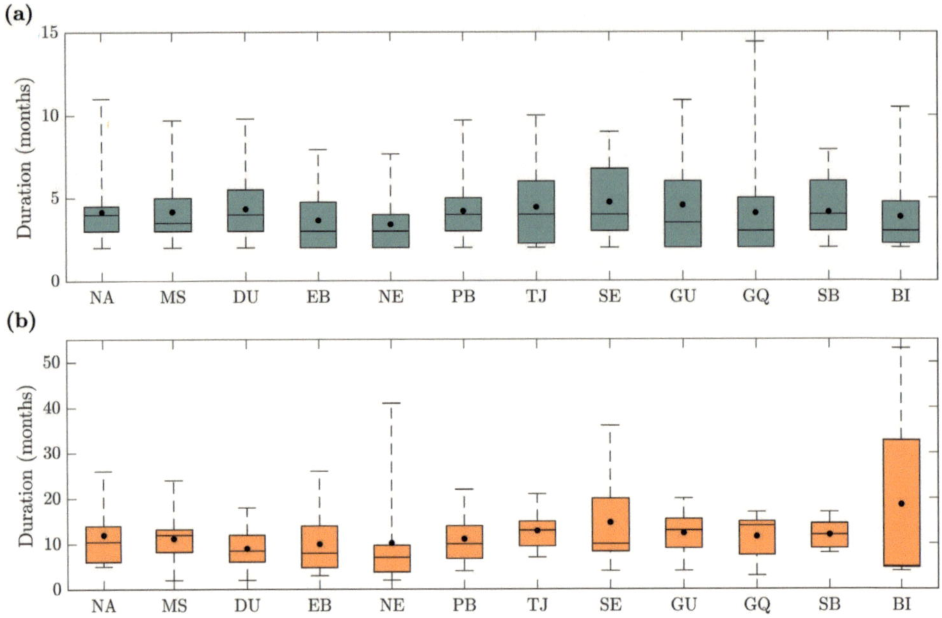

Figure 3.
Box plots of duration of drought events for all river basins, (a) for the SPEI-03 and (b) for the SPEI-12. The lower and upper parts of the boxes represent the 25th and the 75th percentiles, respectively; the line in the middle of each box is the median and the upper and lower whiskers indicate the 2.5th and 97.5th percentiles, respectively. The mean is displayed with black dots.

basin, which presented a large difference between the median and the third quartile (a difference of about 27 months). This latter feature reveals that most of the events for this period were short, but a longer event also occurred. The same behavior also appears in the SE river basins, but here the distribution is more homogenous, showing shorter events.

The box plots for the intensity of the drought events were also examined. In fact, the effects of drought phenomenon are stronger for longer events, but they are also the results of strong intensities. Hence, the distribution of the intensity has to be analyzed. In general, both time scales presented distributions with a low spread, particularly for those events from the SPEI-12. For the 3-month time scale (**Figure 4a**), the median was around 1 for all river basins, indicating that the events were, on average, moderate. As for duration, the mean was slightly higher than the median, reaching values of around 1.1. Concerning the extreme values, the 97.5th percentile was around 1.5 in general and reached a maximum above 1.7 (severe drought) over the EB and the PB river basins. Slightly lower medians are presented for events computed at 12 months (**Figure 4b**), where median intensities rise to values between 0.86 and 1.17 (i.e., from mild to moderate droughts). The mean values were around 1, and these were slightly higher than the median values in most of them. Some basins as MS, EB, or PB presented the opposite behavior. Concerning the extreme values in terms of intensity, the 97.5th percentile was also slightly lower than for the previous time scale (values around 1.45), reaching values above 1.5 over the GQ, the SB, and the PB river basins.

In addition, to examine trends of drought occurrences along the study period, the duration, intensity, and severity of drought events were analyzed by computing such parameters in two different periods: 1980–1999 and 2000–2014. Then, the median changes between both periods in the three parameters were used to develop a categorical classification.

Figure 5 displays the trends for the main river basins and for both time scales. At 3 months (left map), different behaviors were shown. The DU river basin has a positive trend in the three parameters, and the EB presented the opposite behavior. Intermediate trends appeared in the rest of the IP, showing watersheds where a positive trend occurred only in the median severity (i.e., the MS and the SE river basins). In other watersheds, the positive trend only appeared in the median intensity (i.e., the NE, the BI, and GQ river basins), but there were also watersheds with an increase in two of the characteristics analyzed. However, for the 12-month time scale (right

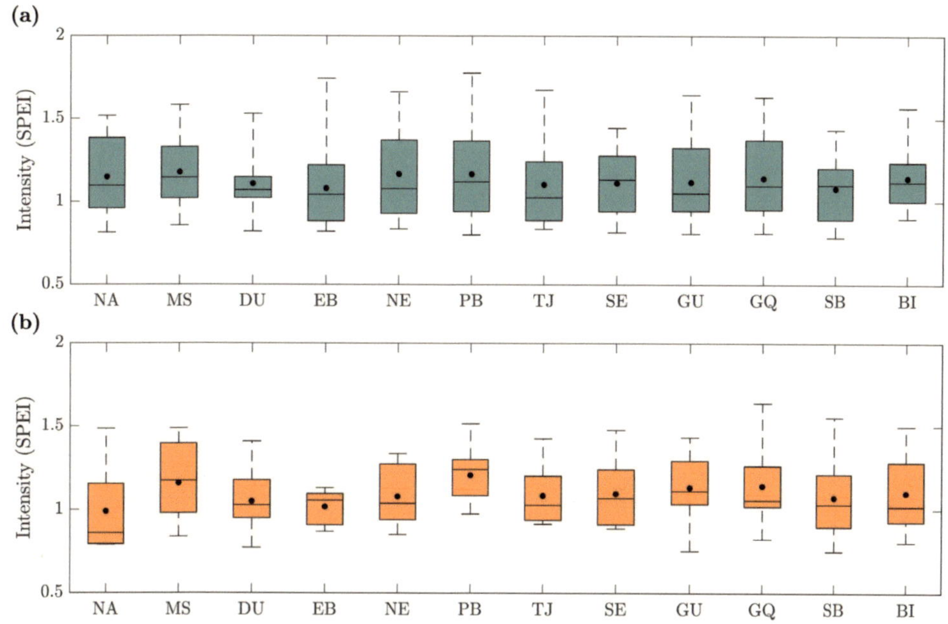

Figure 4.
Box plots of intensity of drought events for all river basins, (a) for the SPEI-03 and (b) for the SPEI-12. The lower and upper parts of the boxes represent the 25th and the 75th percentiles, respectively; the line in the middle of each box is the median and the upper and lower whiskers indicate the 2.5th and 97.5th percentiles, respectively. The mean is displayed with black dots.

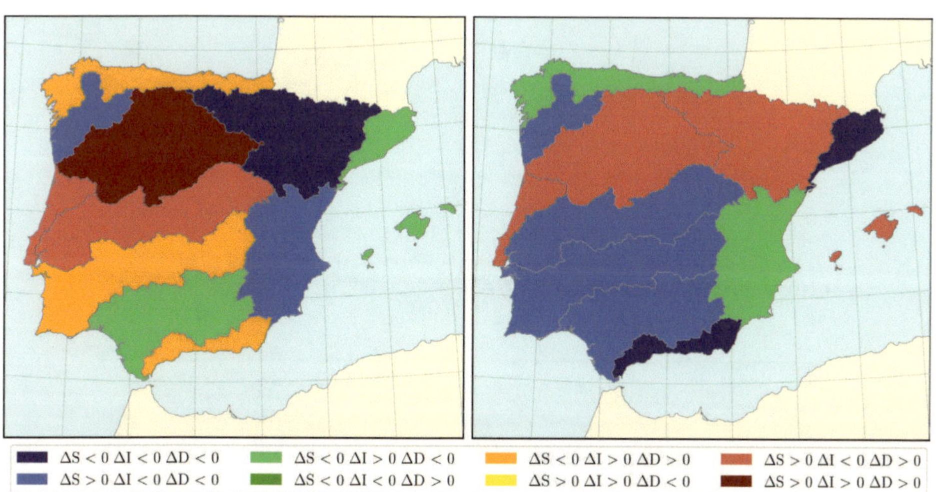

Figure 5.
Categorical classification of droughts for the SPEI-03 (left) and the SPEI-12 (right) based on the changes in severity (ΔS), intensity (ΔI), and duration (ΔD) between the periods 1980–1999 and 2000–2014.

map), the positive trends are less frequent, showing an increased trend in more than one parameter only for the PB, the DU, and the EB river basins. In these three basins, the results showed an increase in both severity and duration of drought events.

4. Temporal evolution of drought-related variables

Finally, we investigated the relationship between the SPEI and other drought-related variables. **Figure 6** shows the temporal evolution of the standardized anomalies of SFCEVP accumulated at 3 months as well as of the monthly mean SM for all river basins. The SPEI-03 was also displayed with comparative purposes. In broad terms, the SM evolves similar to the SPEI-03, showing correlations above 0.7 in all river basins (second column in **Table 3**). However, the SFCEVP presents more

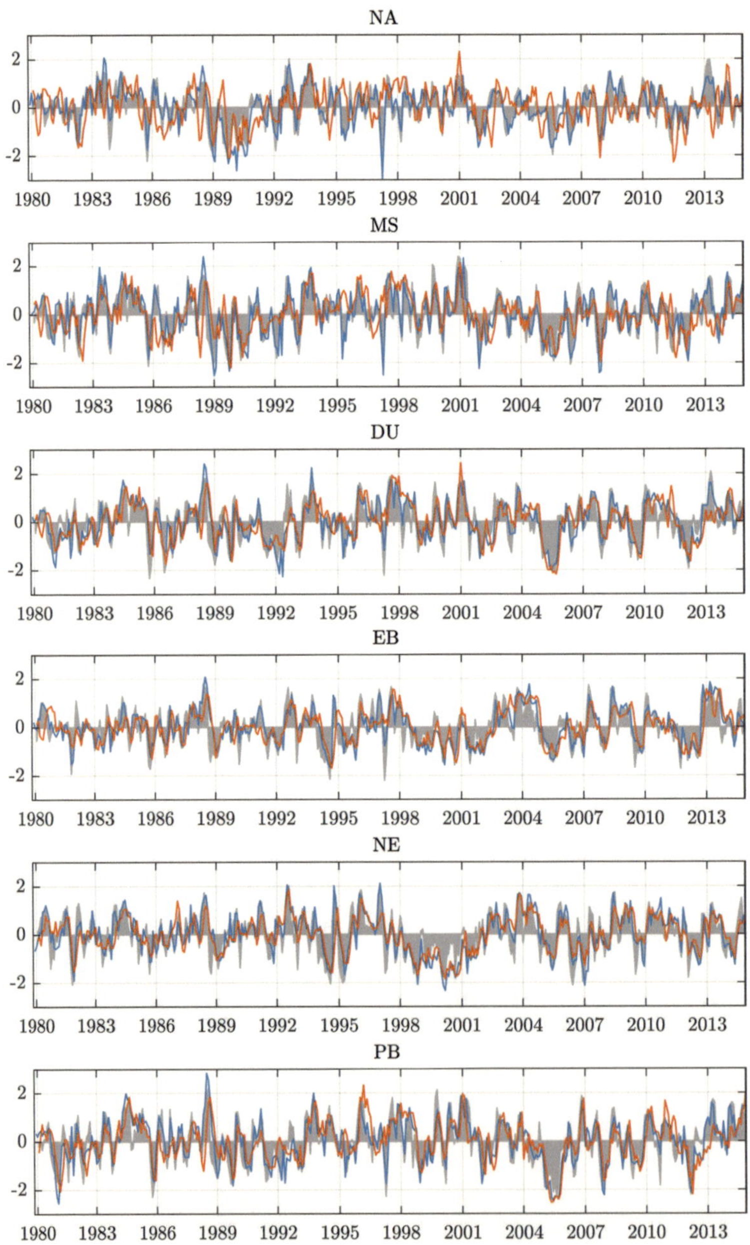

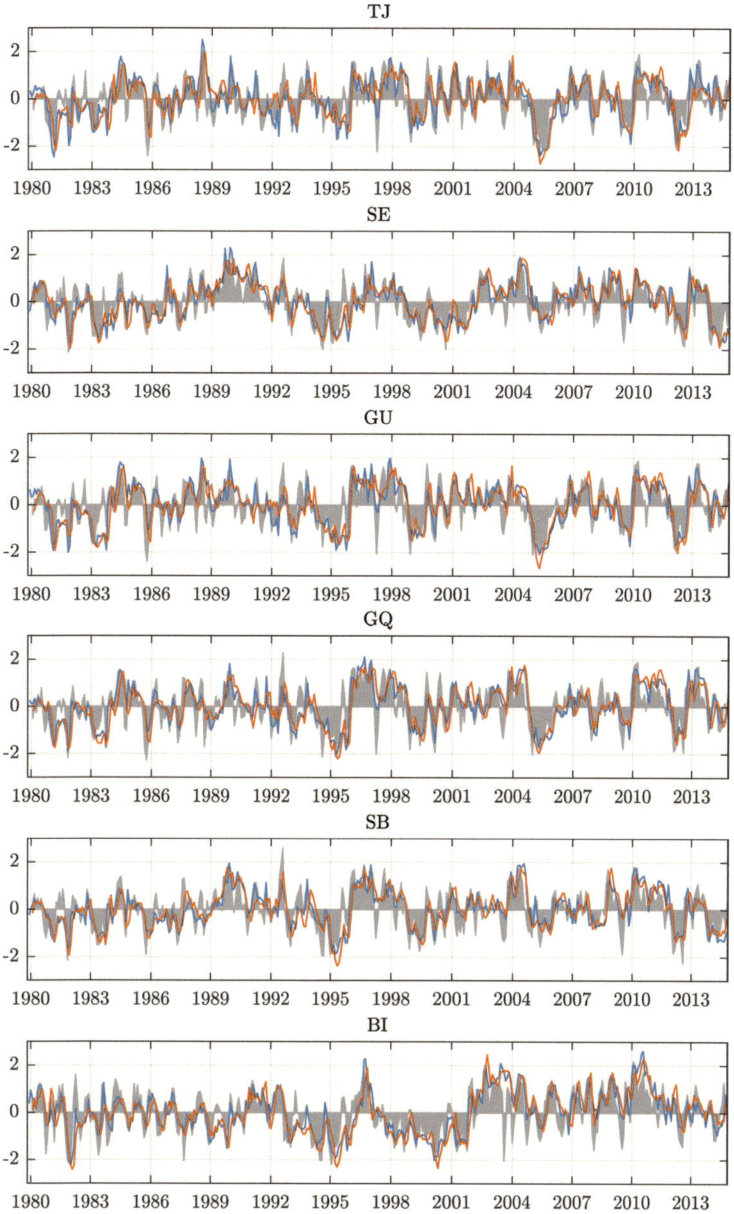

Figure 6.
Standardized anomalies of the 3 months accumulated SM (blue line) and SFCEVP (orange line), SPEI-03 (gray area) evolutions for the NA, MS, DU, EB, NE, PB, TJ, SE, GU, GQ, SB, and BI river basins.

River basins	SPEI-SM	SPEI-SFCEVP	SPEI-runoff
NA	0.85*	0.24	0.91*
MS	0.84*	0.49*	0.91*
DU	0.82*	0.56*	0.80*
EB	0.81*	0.53*	0.80*
NE	0.81*	0.67*	0.80*
PB	0.84*	0.60*	0.87*

River basins	SPEI-SM	SPEI-SFCEVP	SPEI-runoff
TJ	0.79*	0.56*	0.84*
SE	0.80*	0.70*	0.80*
GU	0.76*	0.60*	0.80*
GQ	0.77*	0.61*	0.81*
SB	0.75*	0.67*	0.81*
BI	0.76*	0.63*	0.77*

**Significant correlations at 95% confidence level.*

Table 3.
Pearson's temporal correlation (r) between the SPEI-03 and the soil moisture content (SPEI-SM) and SFCEVP (SPEI-SFCEVP), and between the SPEI-12 and the 12 months' accumulated runoff (SPEI-runoff).

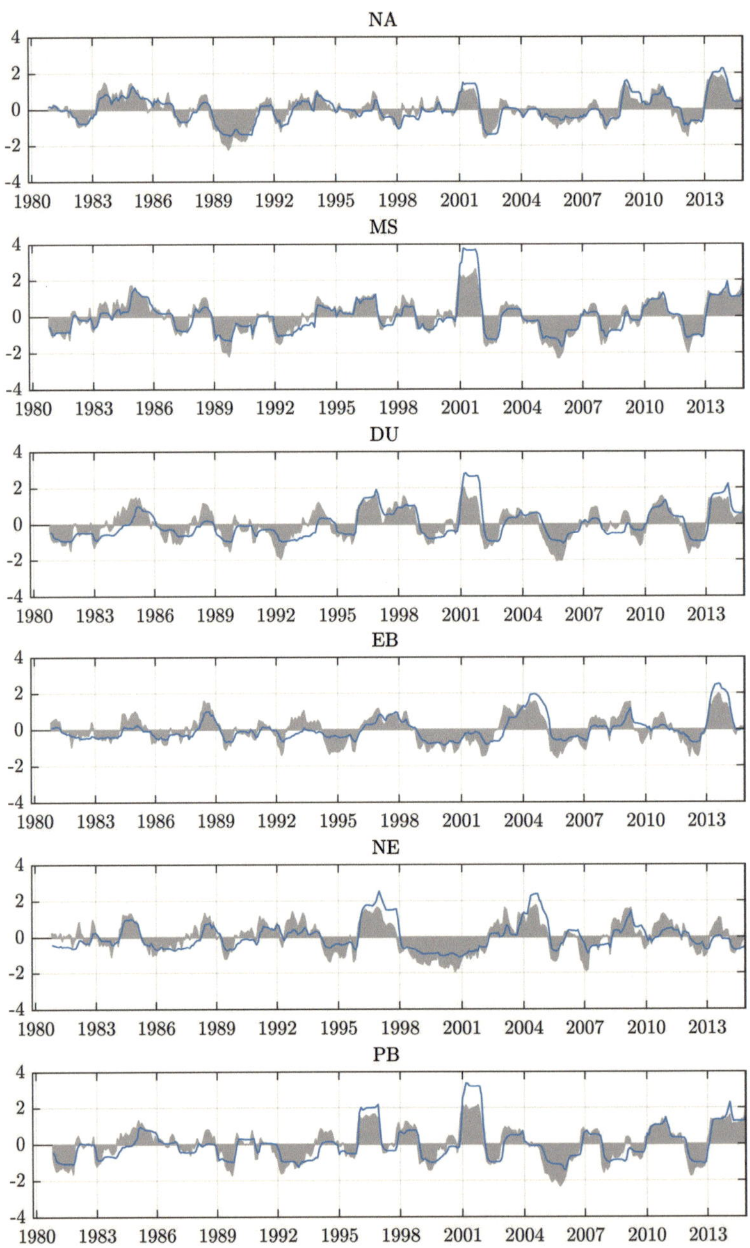

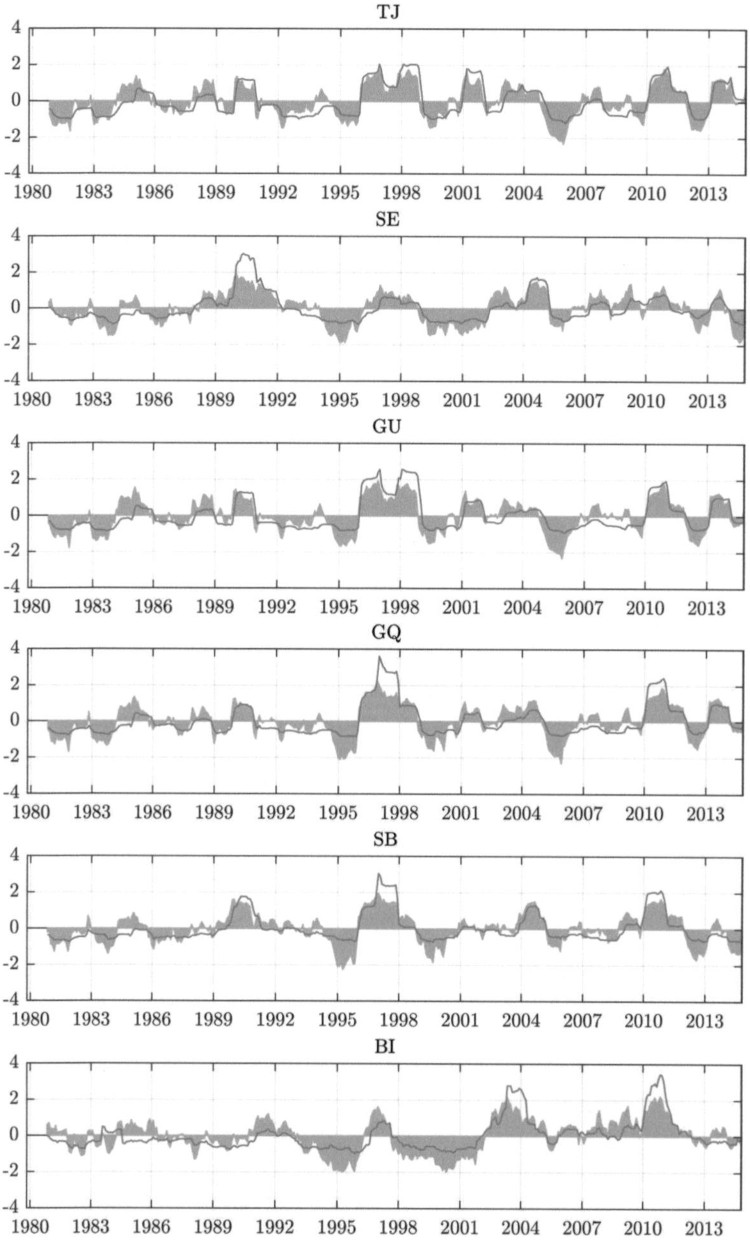

Figure 7.
Standardized anomalies of the 12 months accumulated runoff (blue line) and the SPEI-12 (gray area) temporal series for NA, MS, DU, EB, NE, PB, TJ, SE, GU, GQ, SB, and BI river basins.

discrepancies, especially for the NA and the MS river basins. This latter is probably produced because wet zones such as the northwestern IP are characterized by a SFCEVP less sensitive to changes in SM, being energy-limited regions. Here, only under very long periods of precipitation scarcity, the SFCEVP is strongly affected by the water availability. Contrariwise, in transitional zones (i.e., southern IP), the SFCEVP is usually limited, and then, the SM, and consequently the precipitation, largely controls the SFCEVP. In this region (e.g., the GQ or the SE river basins), the SFCEVP agrees really well with the SPEI-03. This behavior is reflected in the temporal correlations between the SFCEVP and the SPEI-03 (third column in **Table 3**), which tends to be higher over the southern river basins. In broad terms, the Pearson's correlations also show that the SPEI-03 correlates well in general with the SFCEVP, showing significant correlations at the 95% confidence level for all river basins, except over the NA river basins.

The hydrological droughts are represented in **Figure 7**. Here, the temporal evolution of the runoff, which was obtained by aggregated standardized anomalies at 12 months, was displayed, as well as the SPEI-12. As for the previous analysis, different periods with positive and negative anomalies were recorded in all river basins, reproducing wet and dry periods similar to the SPEI-12. In fact, the SPEI-12 reflects long-term precipitation patterns, being thus useful to determine streamflow and reservoir levels. For instance, a marked wet period clearly appeared during 2001 over the NA, the MS, the DU, the PB, and the TJ river basins in both the runoff and the SPEI-12 temporal series. In the same way, certain river basins presented marked negative standardized anomalies of runoff during 2005, which also showed important negative values of SPEI-12. As expected, r between the SPEI-12 and the runoff was really high showing values around 0.85, with the highest ones over the NA and the MS river basins (r values of 0.91).

5. Conclusions

This chapter is devoted to analyzing spatiotemporal patterns of droughts over the IP, a region particularly vulnerable to this extreme event [7] for the period 1980–2014. To do this, a regional climate simulation using the WRF model was completed, obtaining thus high spatiotemporal resolution climate data. Temperature and precipitation data from WRF were used to compute the SPEI at 3- and 12-month time scales with the purpose of analyzing different drought applications (e.g., agricultural and hydrological). This index has proved to be adequate to analyze droughts in the context of global warming.

Firstly, the regional performance of the SPEI was assessed. The most general evolution in droughts is consistent with other studies that allocated the main drought episodes occurred over the IP (e.g., [27]). The results evidenced that WRF appears as a promising tool for analyzing droughts since they provide a great number of climate variables at adequate spatial and temporal resolutions, which are unusually obtained from observations. In fact, climate data from regional simulations may be useful to determine where and when drought is occurring, providing valuable information to compute different drought indices using different drought-related variables.

These results also show that the SPEI-03 represents more frequent and shorter episodes, with the median duration and intensity of about 3–4 months and 1, respectively. In this regard, the largest events occurred over the southern IP, which is especially vulnerable due to its aridity conditions. In fact, this region is arid and then major damages may be produced under the occurrence of severe drought events.

On the other hand, the SPEI-12 presented longer drought episodes, showing, in general, more spread in its duration distributions, with the median values ranging from 5 to 17 months. However, in regard to the intensity, droughts presented similar behavior for both time scales.

Concerning the existence of the trends in droughts along the analyzed period, the results indicate that, in general, an increase in at least one of the drought properties occurred over the period 2000–2014 with respect to the period 1980–1999 in most of the river basins, more evident for the SPEI-03.

In the comparison between the SM and the SPEI-03, the results show that this variable presents a good agreement with the SPEI, which is an index based on a balance between water supply and potential demand at 3 months, which suggests that the 3-month time scale is adequate to study agricultural drought in general over the IP. In fact, it is well known that variations in the SM are the result of changes in precipitation, SFCEVP, and runoff. Therefore, it is an accumulated measure of water availability, and thus, it can be used to detect dry and wet periods [32]. On the other hand, the SFCEVP also affects the agricultural systems, but its relationship with droughts is more complex. That is, the SFCEVP is the actual loss of water, which depends on the water demand by plants and the atmosphere (i.e., the potential evapotranspiration and consequently the temperature), but also on the water availability. This is clearly shown in arid regions, in which the SM largely controls the SFCEVP. Therefore, the actual water balance can be used as a proxy of drought occurrences, but it must be interpreted with caution in regions in which the SFCEVP is not a limiting factor such as the northwestern IP.

Hydrological droughts can be defined as the scarcity of surface and subsurface water resources that leads to negative effects in water resources management system. In this regard, our results indicate that the annual runoff seems to be a good variable to identify hydrological droughts, showing an especial good agreement with the SPEI-12.

The agreement between variables shown here also evidences that the SPEI is an adequate index to investigate drought conditions over the IP, reproducing these in a similar way than other drought-related variables over the study period. This is of high relevance to adequately develop strategies to mitigate the effects of droughts in the future.

Acknowledgements

This study has been financed by the Spanish Ministry of Economy and Competitiveness, with additional support from the European Community fund (FEDER), project CGL2013-48539-R and CGL2017-89836-R. The WRF simulation was run in the ALHAMBRA supercomputer infrastructure (https://alhambra.ugr.es). The ERA-Interim data were obtained from ECMWF portal and the WRF model from NCAR.

Conflict of interest

The authors declare that the research was conducted in absence of any potential conflict of interest.

Author details

Matilde García-Valdecasas Ojeda*, Emilio Romero Jiménez, Sonia R. Gámiz-Fortis, Yolanda Castro-Díez and María Jesús Esteban Parra
Department of Applied Physics, University of Granada, Granada, Spain

*Address all correspondence to: mgvaldecasas@ugr.es

IntechOpen

© 2019 The Author(s). Licensee IntechOpen. This chapter is distributed under the terms of the Creative Commons Attribution License (http://creativecommons.org/licenses/by/3.0), which permits unrestricted use, distribution, and reproduction in any medium, provided the original work is properly cited.

References

[1] Whilite DA. Drought as a natural hazard: Concepts and definitions. In: Whilite DA, editor. Drought: A Global Assessment. London: Routledge; 2010. pp. 3-18

[2] Sheffield J, Wood EF. Projected changes in drought occurrence under future global warming from multi-model, multi-scenario, IPCC AR4 simulations. Climate Dynamics. 2008;**31**:79-105. DOI: 10.1007/s00382-007-0340-z

[3] Dai A. Drought under global warming: A review. Wiley Interdisciplinary Reviews: Climate Change. 2011;**2**:45-65. DOI: 10.1002/wcc.81

[4] Wang K, Dickinson RE, Liang S. Global atmospheric evaporative demand over land from 1973 to 2008. Journal of Climate. 2012;**25**:8353-8361. DOI: 10.1175/JCLI-D-11-00492.1

[5] Vicente-Serrano SM. Spatial and temporal analysis of droughts in the Iberian Peninsula (1910-2010). Hydrological Science Journal. 2006;**51**:83-97. DOI: 10.1623/hysj.51.1.83

[6] Mishra AK, Singh VP. A review of drought concepts. Journal of Hydrology. 2010;**391**:202-216. DOI: 10.1016/j.jhydrol.2010.07.012

[7] Vicente-Serrano SM, Cuadrat-Prats JM. Trends in drought intensity and variability in the middle Ebro valley (NE of the Iberian Peninsula) during the second half of the twentieth century. Theoretical and Applied Climatology. 2007;**88**:247-258. DOI: 10.1007/s00704-006-0236-6

[8] Abatzoglou JT, Barbero R, Wolf JW, Holden ZA. Tracking interannual streamflow variability with drought indices in the US Pacific Northwest. Journal of Hydrometeorology. 2014;**15**:1900-1912. DOI: 10.1175/JHM-D-13-0167.1

[9] Vicente-Serrano SM, Beguería S, López-Moreno JI. A multiscalar drought index sensitive to global warming: The standardized precipitation evapotranspiration index. Journal of Climate. 2010;**23**:1696-1718. DOI: 10.1175/2009JCLI2909.1

[10] McKee TB, Doesken NJ, Kleist J. The relationship of drought frequency and duration to time scales. In: Proceedings of the 8th Conference on Applied Climatology; 17 January 1993. Boston, MA: American Meteorological Society; 1993. pp. 179-183

[11] Vicente-Serrano SM, Beguería S, Lorenzo-Lacruz J, Camarero JJ, López-Moreno JI, Azorin-Molina C, et al. Performance of drought indices for ecological, agricultural, and hydrological applications. Earth Interactions. 2012;**16**:1-27. DOI: 10.1175/2012EI000434.1

[12] Vicente-Serrano SM, Lopez-Moreno JI, Beguería S, Lorenzo-Lacruz J, Sanchez-Lorenzo A, García-Ruiz JM, et al. Evidence of increasing drought severity caused by temperature rise in southern Europe. Environmental Research Letters. 2014;**9**:044001. DOI: 10.1088/1748-9326/9/4/044001

[13] Skamarock WC, Klemp JB, Dudhia J, Gill DO, Barker DM, Duda MG, et al. A Description of the Advanced Research WRF Version 3, NCAR Technical Note. Boulder, Colorado, USA: Mesoscale and Microscale Meteorology Division. National Center for Atmospheric Research; 2008

[14] Dee DP, Uppala SM, Simmons AJ, Berrisford P, Poli P, Kobayashi S, et al. The ERA-interim reanalysis: Configuration and performance of the data assimilation system. Quarterly

Journal of the Royal Meteorological Society. 2011;**137**:553-597. DOI: 10.1002/qj.828

[15] García-Valdecasas Ojeda M, Gámiz-Fortis SR, Castro-Díez Y, Esteban-Parra MJ. Evaluation of WRF capability to detect dry and wet periods in Spain using drought indices. Journal of Geophysical Research: Atmospheres. 2017;**122**:1569-1594. DOI: 10.1002/2016JD025683

[16] Hong SY, Dudhia J, Chen SH. A revised approach to ice microphysical processes for the bulk parameterization of clouds and precipitation. Monthly Weather Review. 2004;**132**:103-120. DOI: 10.1175/1520-0493(2004)132<0103:ARATIM>2.0.CO;2

[17] Betts AK, Miller MJ. A new convective adjustment scheme. Part II: Single column tests using GATE wave, BOMEX, ATEX and arctic air-mass data sets. Quarterly Journal of the Royal Meteorological Society. 1986;**112**: 693-709. DOI: 10.1002/qj.49711247308

[18] Janjić ZI. The step-mountain coordinate: Physical package. Monthly Weather Review. 1990;**118**:1429-1443. DOI: 10.1175/1520-0493(1990)118<1429:TSMCPP>2.0.CO;2

[19] Pleim JE. A combined local and nonlocal closure model for the atmospheric boundary layer. Part I: Model description and testing. Journal of Applied Meteorology and Climatology. 2007;**46**:1383-1395. DOI: 10.1175/JAM2539.1

[20] Chen F, Dudhia J. Coupling an advanced land surface—Hydrology model with the Penn State—NCAR MM5 modeling system. Part I: Model implementation and sensitivity. Monthly Weather Review. 2001;**129**: 569-585. DOI: 10.1175/1520-0493(2001)129<0569:CAALSH>2.0.CO;2

[21] Collins WD, Rasch PJ, Boville BA, Hack JJ, McCaa JR, Williamson DL, et al. Description of the NCAR Community Atmosphere Model (CAM 3.0). NCAR Technical Note NCAR/TN-464+ STR; 2004. p. 226

[22] Kim BS, Chang IG, Sung JH, Han HJ. Projection in future drought hazard of South Korea based on RCP climate change scenario 8.5 using SPEI. Advances in Meteorology. 2016;**2016**(4148710):1-23. DOI: 10.1155/2016/4148710

[23] Wang W, Ertsen MW, Svoboda MD, Hafeez M. Propagation of drought: From meteorological drought to agricultural and hydrological drought. Advances in Meteorology. 2016;**2016**(6547209):1-5. DOI: 10.1155/2016/6547209

[24] Meque A, Abiodun BJ. Simulating the link between ENSO and summer drought in southern Africa using regional climate models. Climate Dynamics. 2015;**44**:1881-1900. DOI: 10.1007/s00382-014-2143-3

[25] Beguería S, Vicente-Serrano SM. SPEI: Calculation of the Standardized Precipitation-Evapotranspiration Index. R Package Version. 2013. p. 1. Available from: http://cran.r-project.org/package=SPEI

[26] Hargreaves GH. Defining and using reference evapotranspiration. Journal of Irrigation and Drainage Engineering. 1994;**120**:1132-1139. DOI: 10.1061/(ASCE)0733-9437(1994)120%3A6(1132)

[27] Beguería S, Vicente-Serrano SM, Reig F, Latorre B. Standardized precipitation evapotranspiration index (SPEI) revisited: Parameter fitting, evapotranspiration models, tools, datasets and drought monitoring. International Journal of Climatology. 2014;**34**:3001-3023. DOI: 10.1002/joc.3887

[28] Allen RG, Pereira LS, Raes D, Smith M. Crop Evapotranspiration-Guidelines for Computing Crop Water Requirements. FAO Irrigation and Drainage Paper 56. Rome: FAO; 1998. 300:D05109

[29] Y-j L, Zheng X-d, Lu F, Ma J. Analysis of drought evolvement characteristics based on standardized precipitation index in the Huaihe River basin. Procedia Engineering. 2012;**28**:434-437. DOI: 10.1016/j.proeng.2012.01.746

[30] Keyantash J, Dracup JA. The quantification of drought: An evaluation of drought indices. Bulletin of the American Meteorological Society. 2002;**83**:1167-1180. DOI: 10.1175/1520-0477-83.8.1167

[31] Bretherton CS, Widmann M, Dymnikov VP, Wallace JM, Bladé I. The effective number of spatial degrees of freedom of a time-varying field. Journal of Climate. 1999;**12**:1990-2009. DOI: 10.1175/1520-0442(1999)012<1990:TENOSD>2.0.CO;2

[32] Sheffield J, Wood EF. Global trends and variability in soil moisture and drought characteristics, 1950-2000, from observations-driven simulations id the terrestrial water cycle. Journal of Climate. 2008;**21**:432-458. DOI: 10.1175/2007JCLI1822.1

Chapter 9

Climate Risk and Early Warning Systems (CREWS) for Papua New Guinea

Yuriy Kuleshov, Kasis Inape, Andrew B. Watkins, Adele Bear-Crozier, Zhi-Weng Chua, Pingping Xie, Takuji Kubota, Tomoko Tashima, Robert Stefanski and Toshiyuki Kurino

Abstract

Developing and least developed countries are particularly vulnerable to the impact of climate change and climate extremes, including drought. In Papua New Guinea (PNG), severe drought caused by the strong El Niño in 2015–2016 affected about 40% of the population, with almost half a million people impacted by food shortages. Recognizing the urgency of enhancing early warning systems to assist vulnerable countries with climate change adaptation, the Climate Risk and Early Warning Systems (CREWS) international initiative has been established. In this chapter, the CREWS-PNG project is described. The CREWS-PNG project aims to develop an improved drought monitoring and early warning system, running operationally through a collaboration between PNG National Weather Services (NWS), the Australian Bureau of Meteorology and the World Meteorological Organization that will enable better strategic decision-making for agriculture, water management, health and other climate-sensitive sectors. It is shown that current dynamical climate models can provide skillful predictions of regional rainfall at least 3 months in advance. Dynamical climate model-based forecast products are disseminated through a range of Web-based information tools. It is demonstrated that seasonal climate prediction is an effective solution to assist governments and local communities with informed decision-making in adaptation to climate variability and change.

Keywords: drought, climate risk, early warning systems, seasonal climate prediction, Papua New Guinea

1. Introduction

Climate has been changing on a global scale since the beginning of the Industrial Revolution; particularly rapid climate change including changes in many extreme weather and climate events has been observed since about the 1950s [1]. The increase in the frequency and severity of extreme weather and climate events has resulted in an increased number of natural disasters impacting on the wellbeing of society. Asia-Pacific is one of the world's most disaster-prone regions, according to the "Asia-Pacific Disaster Report 2012" prepared by the United Nations (UN) Economic and Social Commission for Asia and the Pacific and the UN Office for Disaster Risk Reduction: "almost 2 million people were killed in disasters between 1970 and 2011, representing 75% of all disaster fatalities globally; the most frequent hazards in the region are hydro-meteorological, which affect the most people; since 2000, more than 1.2 billion people have been exposed to hydro-meteorological hazards alone, through 1215 disaster events" [2].

Economic losses in Asia and the Pacific in 1970–2016 attributed to disasters including droughts, floods, storms, earthquakes and tsunamis total to about $1.3 trillion [3]. Economic losses have been rising over the past decades, and it is projected that losses will continue to increase. UN assessment indicates that increase in frequency and magnitude of disasters combined with the increased vulnerability of society could cost the Asia-Pacific region $160 billion per year by 2030 [4].

Developing countries, least developed countries (LDCs) and Small Island Developing States (SIDS) are particularly vulnerable to impact of climate change and climate extremes, including drought, which could lead to water crisis or severe food shortage. For example, a prolonged drought episode related to the strong 2010–2011 La Niña event affected multiple SIDS in the Pacific, including Samoa, Tokelau, Tonga and Tuvalu. The impact of the drought was particularly severe in Tuvalu resulting in water crisis [5]. The Government of Tuvalu declared a state of emergency due to critically low water supplies, and households were rationed to about 40 L of freshwater per family per day [6]. In Papua New Guinea, severe drought caused by strong El Niño event in 2015–2016 affected about 40% of the population, with almost half a million people experiencing food shortages [7].

The Climate Risk and Early Warning Systems international initiative was established in 2015 [8] to enhance early warning systems (EWSs) for vulnerable countries dealing with climate change. CREWS presently operates in countries in Africa, the Pacific and the Caribbean, providing EWS to protect the most vulnerable populations against hydrometeorological hazards like tropical cyclones, droughts and floods [9]. In most of those countries, meteorological observation networks are currently barely adequate, and EWS are basic. The World Meteorological Organization (WMO), working in partnership with the national governments and hydrometeorological agencies of LDCs and SIDS through the projects of the CREWS initiative, strives to improve decision-making around climate change adaptation, disaster risk reduction and sustainable development.

In Africa—in the Democratic Republic of the Congo, Burkina Faso, Mali and Niger—CREWS supports the improvement of operational hydrometeorological forecasts and early warnings for agriculture and food security related to flood and drought risks. In the Caribbean, CREWS assists countries in the region to strengthen regional and national systems and capacity related to weather forecasting, hydrological services, multi-hazard impact-based warnings and service delivery, including tropical cyclones, for enhanced decision-making. In the Pacific—in Fiji, Cook Islands, Kiribati, Niue, Tuvalu, the Federated States of Micronesia, Samoa, Solomon Islands, Tonga, Palau, Nauru, the Marshall Islands, Tokelau and Vanuatu—CREWS projects strengthen the capacity of SIDS in hydrometeorological services and EWS. In Papua New Guinea, CREWS improves the existing drought monitoring network, as well as early warnings for the agriculture sector and emergency service managers.

In this chapter, the CREWS project for PNG and its implementation strategy are described; synergies with the International Climate Change Adaptation Initiative, the PNG Capacity Development Program and the Space-Based Weather and Climate Extremes Monitoring Demonstration Project (SEMDP) are outlined; and the project's preliminary results are presented.

2. CREWS-PNG

The CREWS-PNG project develops improved drought monitoring and subseasonal-to-seasonal prediction that can foster better decision-making for agriculture, water management and other climate-sensitive sectors by creating an end-to-end EWS aimed at reducing the impacts of drought. In this section, a brief overview of climate impacts on PNG climate-sensitive sectors is presented, providing the rational for CREWS-PNG project and its implementation strategy.

2.1 Climate impacts on PNG climate-sensitive sectors

PNG is a country in the Southwest Pacific with a population of almost 7 million people, with agriculture providing a subsistence livelihood for 85% of the population. PNG, the largest of the Pacific Island Countries, faces multiple climate change-related challenges. Climate-related natural disasters, as well as gradual shifts in climatic and oceanic conditions, already pose significant risks to PNG, disrupting daily life, causing damage to assets and infrastructure, destroying livelihoods and killing or injuring people.

The El Niño-Southern Oscillation (ENSO) is one of the key drivers of interannual climate variability around the globe, with substantial impacts felt in PNG. During the warm phase of ENSO (El Niño), ocean surface waters in the central and western Pacific Ocean cool, causing a shift in weather patterns and associated rainfall towards the eastern Pacific. This often results in severe rainfall deficits in the western Pacific, leading to significant drought conditions across PNG. However during the cool phase of ENSO (La Niña), warmer than average ocean surface waters and enhanced convection and rainfall occur over the Maritime Continent, which can result in extreme precipitation, flooding and landslides in PNG. These climate extremes take a severe toll on both the population and economy of PNG people. For instance, the 2015–2016 El Niño, which was one of the three strongest El Niño events since the 1950s, had significant impacts on PNG agriculture and energy and mining sectors.

For agriculture, the 2015–2016 El Niño-induced drought and frost led to crop failures which affected almost 2.5 million people (approximately 40% of the population) with almost half a million people suffering severe food shortages [7]. Staple sweet potato crops in the highlands were severely damaged by frosts in August 2015—the result of reduced night-time cloud cover—which also destroyed wild plants that are usually eaten as a supplement source of food.

For energy, the hydropower plant at Yonki Dam, which supplies the entire highlands and the Momase regions with electricity, and the Sirinumu Dam, which supplies the nation's capital Port Moresby, were not able to adequately meet the expected energy demands due to low water levels. As a result, energy suppliers were forced to switch to diesel power generation which was both costly and environmentally unfriendly.

For mining, Ok Tedi, one of PNG's major mining operations, closed down its operations due to very low water levels in the Fly River which prevented cooper ore and other minerals from being shipped out.

It is expected that climate change will exacerbate those hazards already impacting on agricultural yields and the productivity of other economy sectors, further reducing the financial and social wellbeing of PNG population. Addressing these issues, the government of PNG through the Office of Climate Change and Development has put its emphasis on combating natural hazards in the country, including food insecurity caused by crop failures due to droughts and inland frosts.

The CREWS-PNG project aims to address this government priority through enhancing EWS and strengthening resilience to climate change by providing accurate and timely information about the current state of the climate and increasing the use of subseasonal-to-seasonal climate predictions. This will have a primary focus on drought- and food security-related impacts while providing benefits in other priority areas, e.g. climate change mitigation and adaptation, water and sanitation impacts of drought, etc.

2.2 CREWS-PNG key areas of activity

The main objective of the CREWS-PNG project is to build the capacity of the PNG National Weather Service and strengthen its cooperation with key business sectors, government departments and other stakeholders, in order to put in place complete systems that deliver warnings and relevant information to end-users. The following key components of the project—improvement of observations and databases, weather and climate monitoring and forecasts and support for EWS development—are designed in order to achieve this objective.

2.2.1 Improvement of observations and databases

Meteorological observations (i.e. collecting instrumental data about atmospheric pressure, temperature, moisture, precipitations, wind, solar radiation, ocean temperature and many other weather and climate essential variables which describe state of the environment) are one of the mandatory functions of National Meteorological and Hydrological Services (NMHSs) around the world. The PNG NWS currently has an observation network of 13 weather and climate stations, 7 rain gauge stations and 5 agrometeorological stations.

Given PNG's total area is about 462,840 sq. km with rugged terrain and many remote microclimates, the existing meteorological observation network is not adequate to provide a comprehensive spatial coverage. Improvement of meteorological observations is an essential part of the project and is addressed through increased station network reliability and sustainability and improved control and maintenance procedures as well as through augmenting data from the existing surface-based observation stations with the data from additional automatic weather stations (AWSs) that will be established in partnership with the PNG Capacity Development Program (PNG-CDP) (see Section 2.3 for detail).

Collecting meteorological observations over a substantial period of time (e.g. the WMO standard for deriving climatology is 30 years of records) is vital for understanding the current climatic conditions as well as deriving trends in environmental variables. In some countries, NMHSs have long-term climate records going back to the nineteenth century. Historically, meteorological observations were kept as paper records. In modern times, these paper records have been digitized and stored electronically in computerized database management systems, and new data are archived in digital form.

In PNG, early historical records of rainfall for Port Moresby are available from the 1890s and are largely complete from 1905. Rainfall data from other meteorological stations located across the country are also available, typically from the 1950s to 1960s. Some historical climate records have been digitized as part of the International Climate Change Adaptation Initiative (ICCAI). However, substantial climate records are still stored in paper form and required digitization. One of the key areas of activities for CREWS-PNG is to continue the implementation of climate data rescue activities. Data digitizing is conducted at the climate section of the PNG NWS under supervision of experienced climatologists. Quality control procedures are in place to ensure that this invaluable historical climate record is preserved for future analysis. The improvement of historical databases, including data homogenization, is essential for deriving accurate climatology and trends. This important task is addressed by the CREWS-PNG project.

Once digitized, the records are archived in a modern climate database management system Climate Data for the Environment (CliDE) which has been used by the PNG NWS operationally for almost a decade. CliDE was developed as part of the ICCAI and implemented at 15 NMHSs across the Pacific, including PNG. CliDE is one of the WMO recommended climate database management systems and is included in the Climate Services Toolkit. CliDE is undergoing continuous development; further tailoring of the database to better suit requirements of the PNG NWS is undertaken as part of this project.

2.2.2 Weather and climate monitoring and forecasts

Currently, nationwide weather forecasts issued by the PNG NWS and provided to the media have lead times of 24 h, while seasonal climate predictions are produced at longer lead times of 3 months and limited to site-specific locations. While weather forecasts and seasonal climate predictions are currently recognized by users as useful products produced by PNG NWS, stakeholders from various economic sectors have noted that there is a need to enhance and expand these services.

To enhance the availability of numerical weather prediction (NWP) products, the CREWS-PNG project is increasing the availability of NWP limited area model guidance from the WMO Regional Specialised Meteorological Centre (RSMC) for use in short-range weather forecasts (up to 7 days).

Another valuable improvement of climate services is the transition to seamless forecasting, i.e. providing forecasts and early warnings on time scales from days right to weeks and months. For sub-seasonal to seasonal forecasts (i.e. at lead times of 1–4 weeks and 1–3 months), products from the WMO Global Producing Centres for Long-Range Forecasts (GPCLRFs) and Regional Climate Centres (RCCs) are used.

Development of an objective seasonal forecasting scheme for PNG, with skill measures that will be communicated to users accompanying the forecasts, is underway. This is partly based on the WMO Sub-Seasonal to Seasonal Project to provide daily forecasts for the next 60–90 days. National forecasts will add value to, and be based on, outputs from an objective regional forecasting scheme for the region. The regional seasonal forecasting model will similarly benefit other countries in the Asia-Pacific region.

2.2.3 Support for EWS development

While EWSs are currently in place, they have limited capacity for providing disaster- and food security-related warnings that can trigger effective action. To address this issue, this project will initiate the following activities focused on developing improved products for drought monitoring and prediction.

Currently, PNG NWS provides drought monitoring based on the analysis of rainfall observations from surface-based rain gauge stations, which currently consists of an observation network of 13 weather and climate stations and 7 rain gauge stations. This number of stations is considered inadequate to accurately capture the complex spatial distribution and variability of rainfall across the country.

A modern-day technology—satellite remote sensing—is an approach which will be used by CREWS-PNG to improve drought monitoring for PNG. Space-based observations not only provide global coverage but are now used by NMHSs around the world for weather monitoring and input to weather forecasting models. Space-based observations are a particularly valuable source of information for countries with a low density of surface-based meteorological stations like PNG. Recently, WMO established the SEMDP to assist NMHSs in the Asia-Pacific region with precipitation monitoring (for detail, see Section 2.3). SEMDP drought monitoring products available from NOAA and JAXA will be used by the PNG NWS to enhance its service.

In terms of drought forecasts, the PNG NWS currently produces outlooks based on a statistical model (Seasonal Climate Outlooks in Pacific Island Countries (SCOPIC)), which needs to be updated.

SCOPIC is a seasonal climate prediction tool used operationally by NMHS in 15 Pacific Island Countries for more than a decade [10]. In the past, statistical models demonstrated skill in predicting seasonal rainfall in Australia [11] and countries in the Pacific [10]. However, statistical models are based on long-term historical records and developed under the assumption that climate is stationary. A hindcast validation study which examined the forecast skill of SCOPIC in the Pacific revealed that the model demonstrated reasonable skill using a number of selected predictors, but the model's skill varied depending on the location and the season (e.g. the model's highest rainfall predictive skill was for the austral summer and lowest for the austral winter) [10].

In a warming climate, the underlying assumption about the stationarity of climate is no longer valid, and skill of statistical models deteriorates. Modern seasonal climate prediction models use dynamical climate modelling techniques, which are based on the laws of physics and account for natural climate variability and climate change not captured by historical data. Under the CREWS-PNG project, outputs of state-of-the-art dynamical climate model are used to strengthen the capability of the PNG NWS to perform skilful drought forecasts (see Section 3 for detail).

Based on improved drought monitoring and prediction capabilities at the PNG NWS, the national drought EWS will be enhanced. Drought EWS weather and climate information will be translated into drought early warning alerts and advisories for various sectors, with these trigger points set by a series of stakeholder consultations. Overall, the key project activities aim to provide impact-based forecasts and risk-informed warnings to assist decision-making by the users. To reach "the last mile" and ensure that no one is left behind, multichannel weather forecast and warning communication systems will be developed or enhanced, providing communities with actionable information and alerts.

2.3 Partnership with SEMDP and PNG-CDP

CREWS-PNG is implemented in partnership with the SEMDP and the PNG-CDP, to assist the PNG NWS with enhancing drought monitoring, observation network and weather forecasting. Synergies between the projects are briefly outlined in this section.

2.3.1 SEMDP

Recognizing a need to assist members to better utilize and improve their ability to monitor extreme weather and climate events from space, WMO established the SEMDP—a demonstration project for space-based weather and climate extremes monitoring, with focus on operational monitoring of extreme precipitation, notably drought and accumulated rainfall over 5–7 days [12]. In the first stage of the demonstration project (2018–2019), drought monitoring products will be available from satellite providers NOAA and JAXA for the Asia-Pacific region (SEMDP geographical domain is between 40°N–45°S and 50°E–160°W), with future expansion of the domain to other WMO regions and eventually globally, at subsequent stages of the project. The availability of SEMDP satellite-derived data and products over the Asia-Pacific domain allows them to be utilized by CREWS-PNG, through the PNG NWS, to provide an enhanced drought monitoring service for PNG.

NOAA satellite products for the SEMDP (gauge-based, satellite-estimated and gauge-satellite blended products of precipitation) are produced at the Climate Prediction Center (CPC). To construct a high-quality, high-resolution precipitation analysis over the globe through integrating information from satellite observations as well as in situ measurements and model simulations, the CPC morphing technique—CMORPH—is used (see [13] for detail). CMORPH products start from 1998 and are updated by CPC/NOAA on a quasi-real-time basis. JAXA products for the SEMDP are produced at the Earth Observation Research Centre (EORC). They are based on Global Satellite Mapping of Precipitation (GSMaP) data [14], with GSMaP version 6 (GSMaP_GNRT6) data reprocessed from March 2000.

The SEMDP products available from CPC/NOAA and EORC/JAXA include precipitation estimates at various time scales, e.g. hourly, daily, pentad (5 days), weekly, 10 days and monthly, as well as derived products, e.g. the standardized precipitation index (SPI), the normalized difference vegetation index (NDVI), the vegetation health index (VHI), etc. Here we present a few examples of these SEMDP products in the context of the CREWS-PNG project.

Prior to CREWS-PNG, the climate service team of the PNG NWS used only surface-based station rainfall data and statistics based on rain gauge observations. Utilizing SEMDP products, maps of precipitation on various time scales are now available to assist the PNG NWS with spatial analysis of precipitation. In **Figure 1**, a map of monthly average precipitation over the SEMDP domain for July 2018, derived from CPC/NOAA CMORPH data, is presented. Examining closely the PNG area, it can be seen that in some parts of PNG, only 2 mm/day or less of rain was received in that month compared to the long-term July climatological mean of 5–10 mm/day (**Figure 2**). Another valuable SEMDP product for operational rainfall monitoring is the weekly rainfall statistics, which includes both precipitation anomaly and percentage of normal rainfall (**Figure 3**).

The standardized precipitation index is widely used to characterize and monitor meteorological drought, i.e. drought that occurs when below average precipitation occurs for a prolonged period of time [15]. Standardized precipitation is the

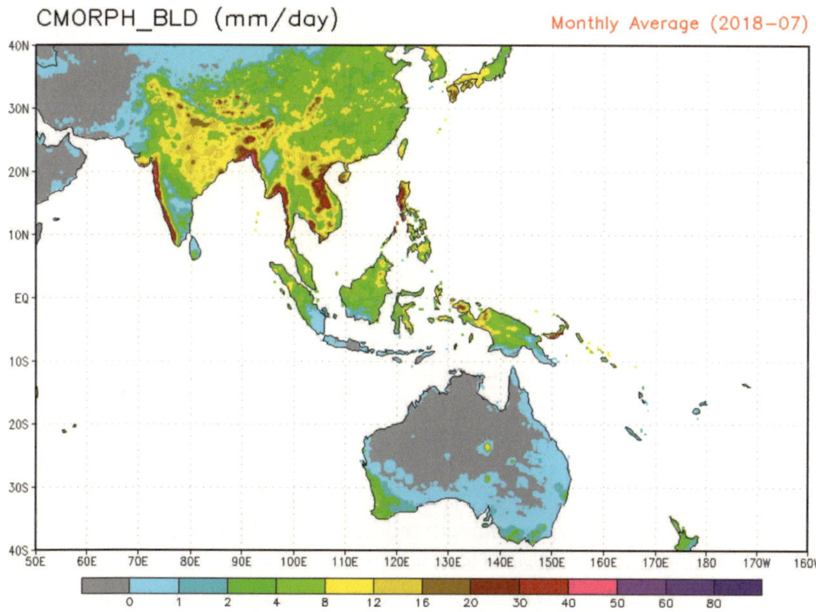

Figure 1.
CPC/NOAA CMORPH monthly average precipitation over the SEMDP domain for July 2018.

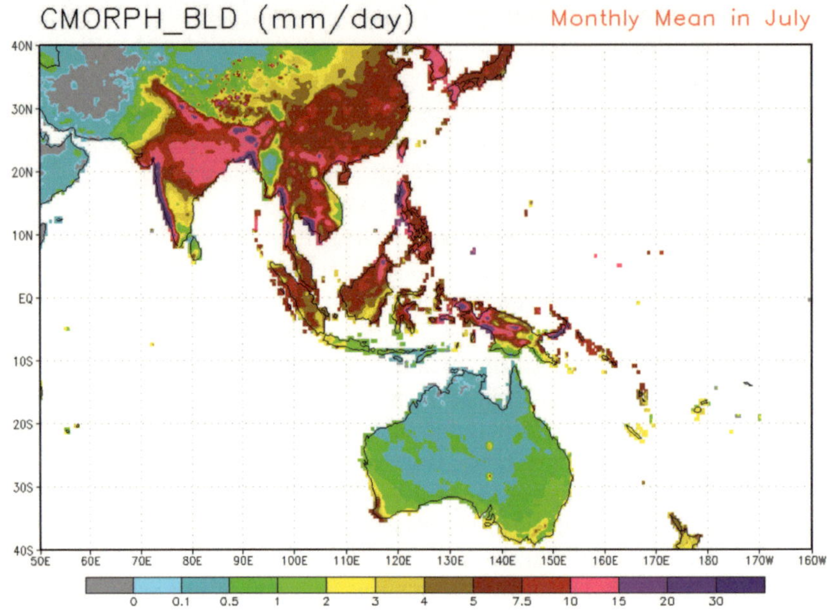

Figure 2.
CPC/NOAA CMORPH monthly mean precipitation (climatology) for July (1998–2017 base period).

difference of precipitation from the mean for a specified time (e.g. 30, 60 and 90 days or longer time period) divided by the standard deviation; the mean and the standard deviation are both determined from the climatological record. The SPI values are interpreted as follows: 2.0 < SPI, extremely wet; 1.5 < SPI < 2.0, very wet; 1.0 < SPI < 1.5, moderately wet; −1.0 < SPI <1.0, near normal; −1.5 < SPI < −1.0, moderately dry; −2.0 < SPI < −1.5, severely dry; and SPI < −2.0, extremely dry. The 60-day and 90-day SPI ending at 2 December 2018 is presented in **Figure 4**, top and bottom panels respectively, indicating severely dry (−1.5 and below) and extremely dry (−2.0 and below) conditions in most parts of the eastern and southern provinces of PNG during September, October and November 2018.

The impact of the 2015–2016 El Niño on the agricultural sector of PNG was discussed earlier. Here we demonstrate the value of space-based observations for drought monitoring utilizing the vegetation health index [16]. The VHI is computed using observations from Advanced Very High Resolution Radiometer (AVHRR), instrument onboard the NOAA polar orbiting satellites, in the visible, infrared and near-infrared bands, and used to identify stress on vegetation related to drought. Maps of the VHI for the first week of January 2015 and the first week of December 2015 are presented in **Figure 5**, top and bottom panels respectively. These maps clearly demonstrate the difference between relatively healthy vegetation over PNG in January and stressed vegetation in December when the impact of El Niño was very strong [the 3-month average of the Oceanic Niño Index (3-month running mean of sea surface temperature anomalies in Niño 3.4 region; for detail, see [17]) from November 2015 to January 2016 had peaked at +2.6°C. Overall, the 2015–2016 El Niño event was classified as one of the three strongest events since the 1950s].

In addition to the impact on vegetation due to drought caused in PNG by the El Niño event in 2015, frost also occurred in highlands. Frost was caused by the reduced night-time cloud coverage allowing increased loss of heat to space during the night, resulting in severe damage to the staple sweet potato crops leading to food shortages. Cloud coverage could be estimated using AVHRR space-based observations to measure the amount of outgoing longwave radiation (OLR). Negative OLR

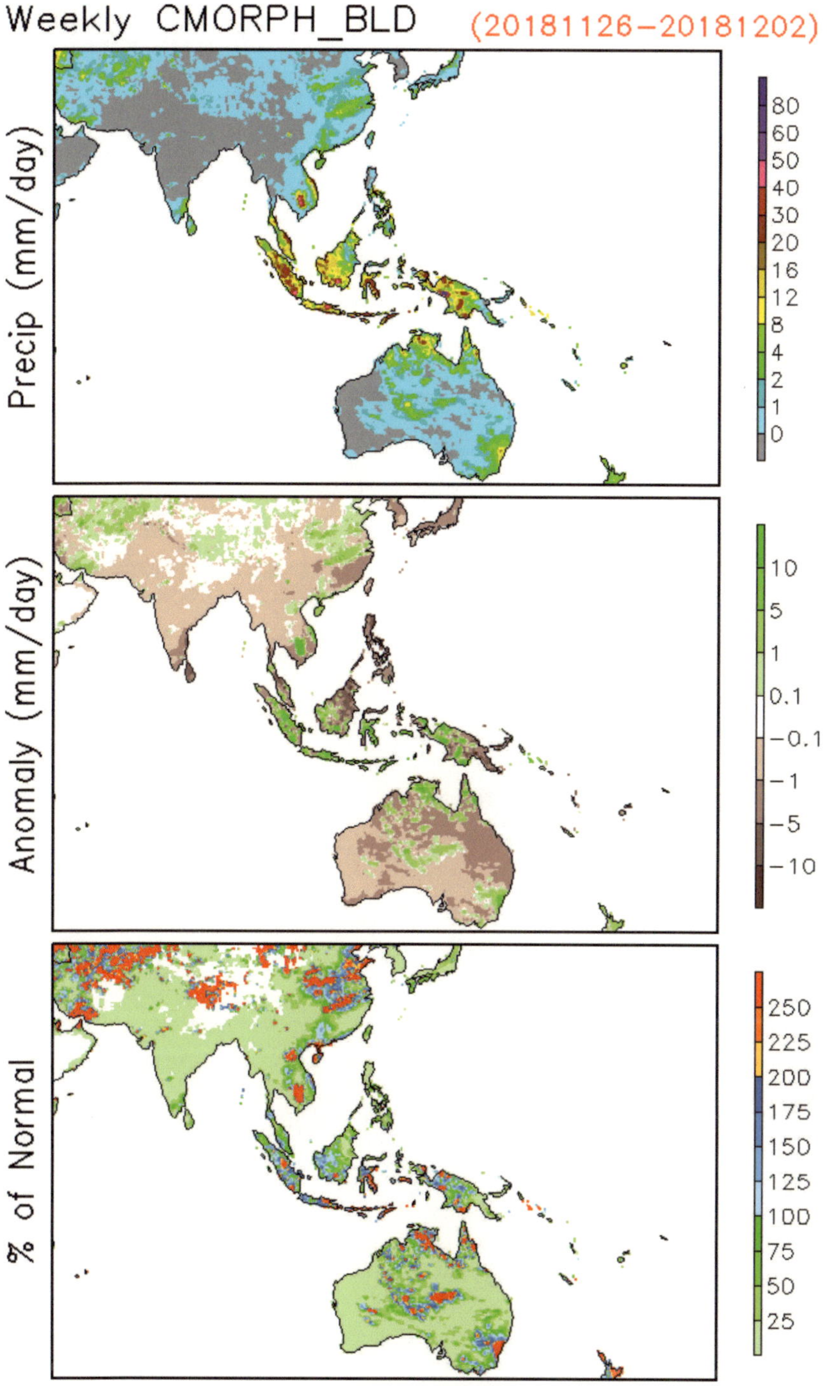

Figure 3.
CPC/NOAA CMORPH weekly summary: mean precipitation, mm/day (top panel); precipitation anomaly, mm/day (middle panel); and precipitation percentage of normal, % (bottom panel) for a week from November 26, 2018 to December 2, 2018.

anomalies indicate enhanced convection and hence increased cloudiness, while positive OLR anomalies indicate suppressed convection and decreased cloudiness. In **Figures 6** and 7, maps of AVHRR OLR and OLR anomalies for 6–10 January 2015 (pentad 2) and 15–19 August 2015 (pentad 45), respectively, are presented. Large

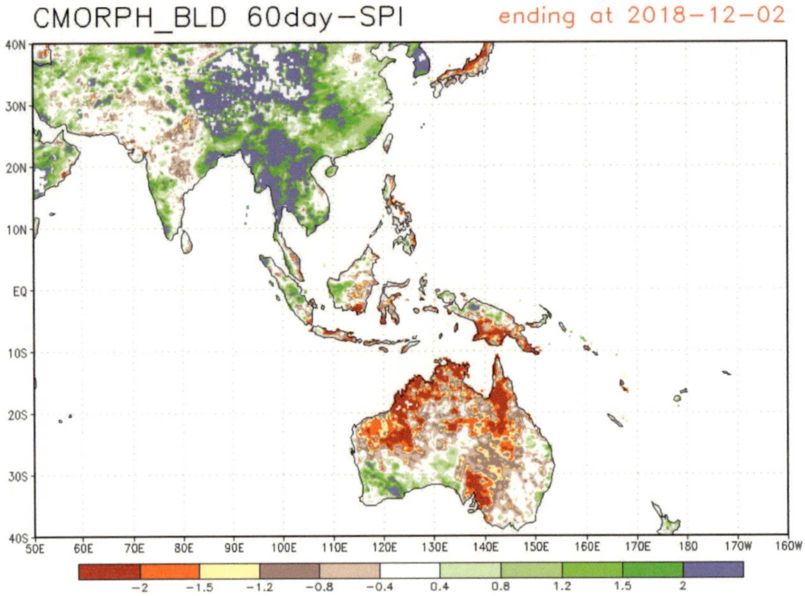

a

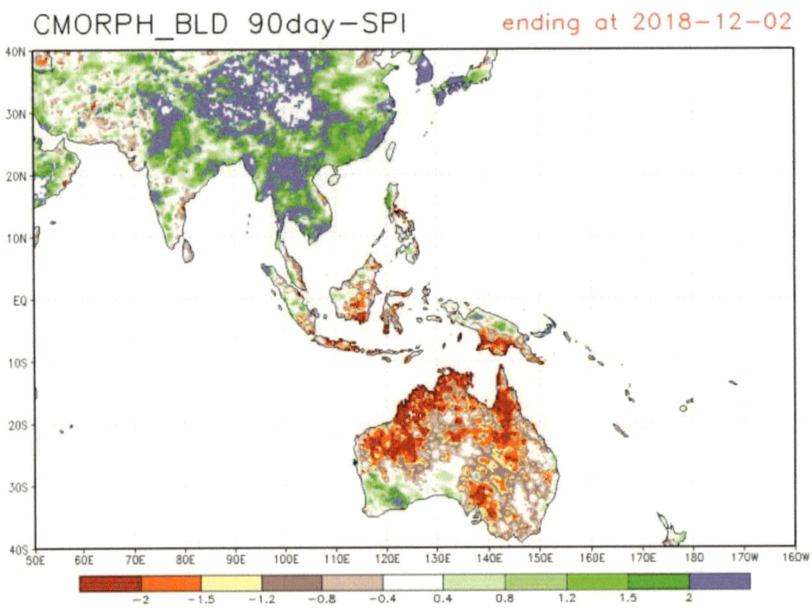

b

Figure 4.
CPC/NOAA CMORPH 60-day (top panel) and 90-day (bottom panel) SPI ending at December 2, 2018.

negative OLR anomaly is evident over PNG in January 2015 indicating more cloud coverage (**Figure 6**, bottom panel), while in August 2015 it is the opposite—OLR anomaly is strongly positive indicating reduced cloud coverage over the country (**Figure 7**, bottom panel).

In summary, the introduction of SEMDP drought monitoring products is significantly strengthening the capacity of the PNG NWS to provide users in agriculture and other sectors with valuable information for decision-making which was not available prior to implementing CREWS-PNG project.

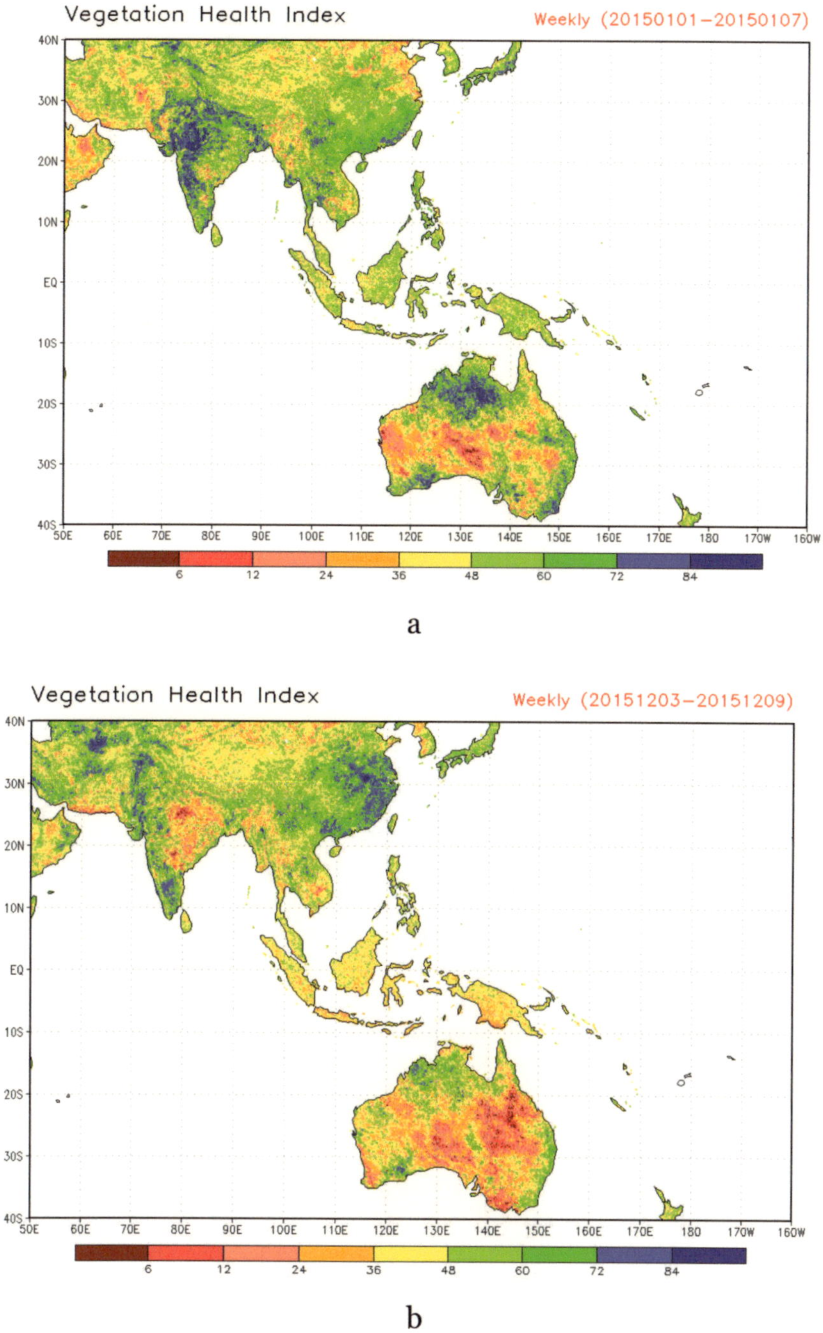

Figure 5.
CPC/NOAA CMORPH VHI for January 1–7, 2015 (top panel) and December 3–9, 2015 (bottom panel).

2.3.2 PNG-CDP

The CREWS-PNG project partners with the PNG-CDP—a project which is focused on enhancing the meteorological observation network in the country. Meteorological observations are critical for the operations of national meteorological services of all 192 WMO Members. The quality of meteorological observations from each country directly affects the accuracy of national as well as global weather forecasts, and the free exchange of these observations and other relevant data is

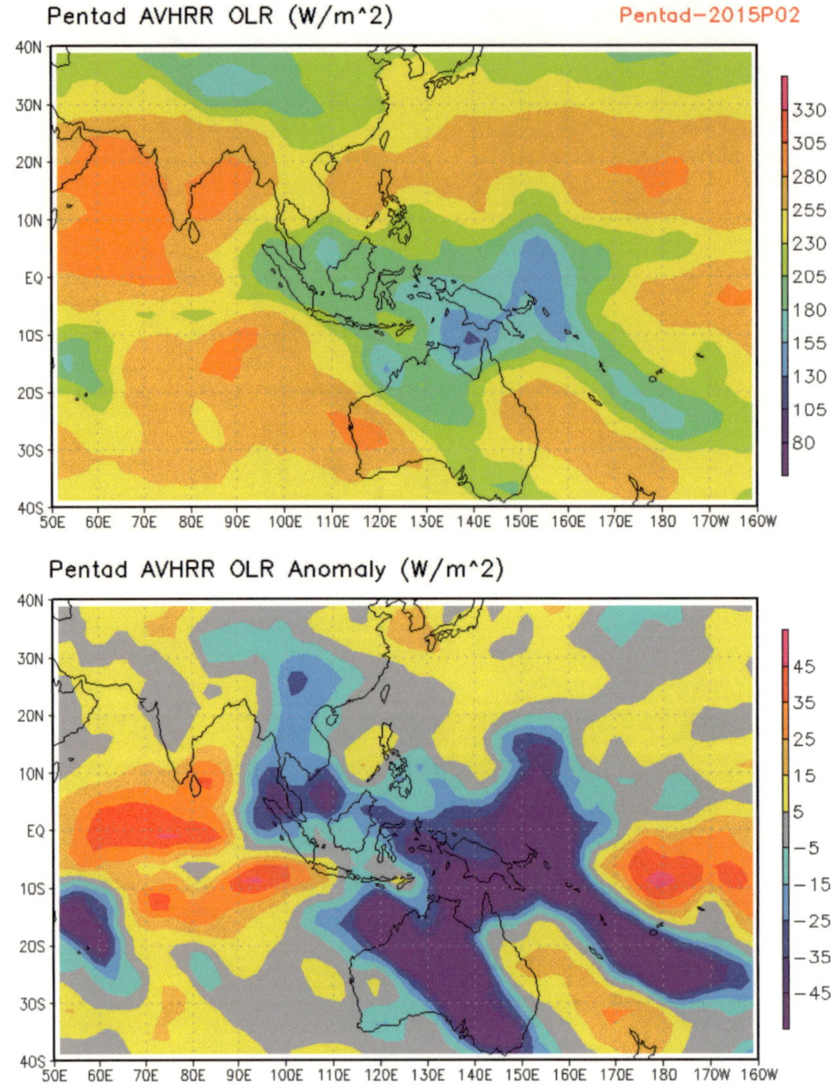

Figure 6.
AVHRR OLR for January 6–10, 2015, i.e. pentad 2 (top panel) and pentad OLR anomaly (bottom panel), W/m².

critical to each forecasting service. The international nature of many sectors, e.g. transport sector (aviation, shipping, road transport, etc.) and tourism, means that forecasts and warnings issued by any country can directly affect the lives of citizens of other countries and the security environment of each region is also influenced by how well weather-related natural disasters are warned for and managed. It is therefore in the global interest that all NMHSs are able to participate fully in the international effort to provide high-quality meteorological observations, weather and climate monitoring and prediction services.

A high functioning weather service is imperative to the transport sector in any country, to allow transport agencies such as air traffic management, maritime regulatory agencies and search and rescue agencies to provide a reliable service to the public. PNG is a nation with intense transport challenges. The topography, multi-island nature, population make-up and tropical climate pose issues for aviation, land and maritime transport. All transport sectors are vulnerable to weather. PNG has a monsoonal climate with intense summer rainfall and a large variation in climate across the country.

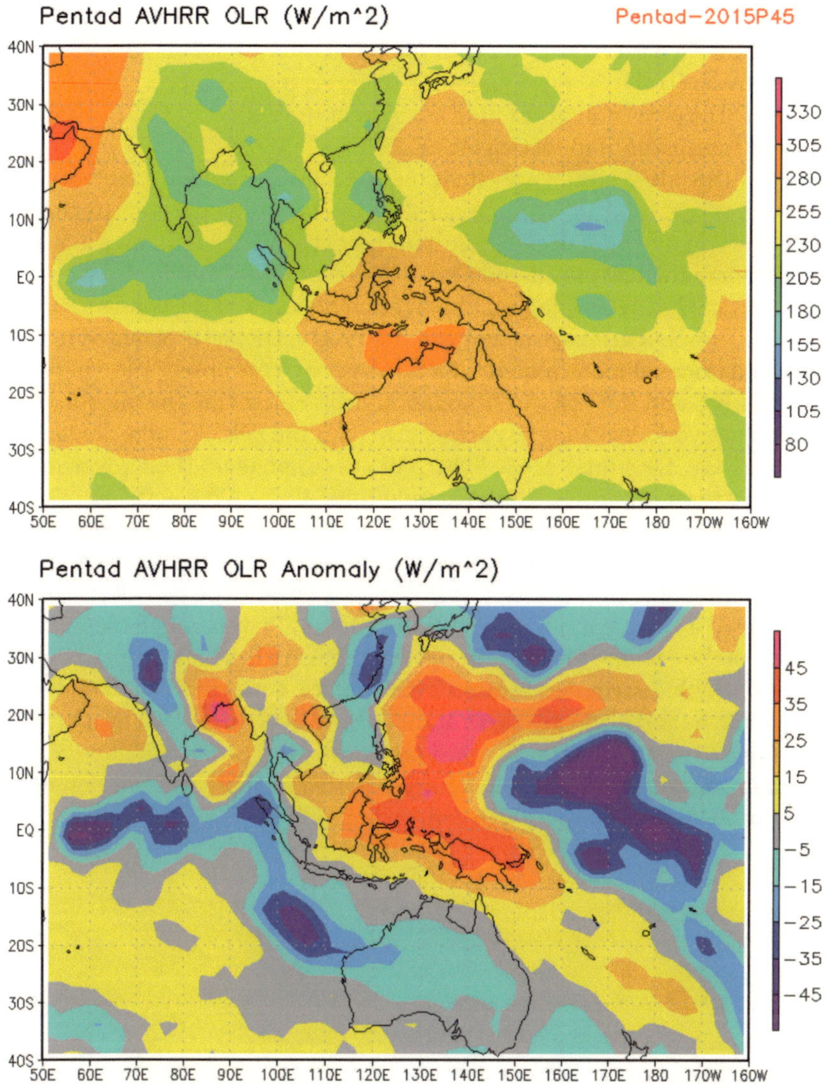

Figure 7.
AVHRR OLR for August 15–19, 2015, i.e. pentad 45 (top panel) and pentad OLR anomaly (bottom panel), W/m².

The PNG NWS is positioned within the Technical Services Division of the PNG Department of Transport (DoT) and operates within a highly challenging physical and social environment, providing critical warning, weather and climate services to a diverse community of marine, aviation and road transport users, emergency service organizations, agricultural users and the general public.

In 2014 the PNG DoT, supported by the Australian Government Transport Sector Support Program (TSSP) and the Australian Department of Infrastructure, Regional Development and Cities (DIRDC), commissioned a "mini-diagnostic" of the PNG NWS. The Australian Bureau of Meteorology and the International Civil Aviation Organization contributed subject matter experts to conduct this assessment. The mini-diagnostic concluded that the current capacity of the PNG NWS is not sufficient to provide high-quality observations and forecasts and there is a need to enhance capacity of PNG weather and climate services.

In response to the findings and recommendations of the mini-diagnostic, PNG and Australia agreed to add meteorology as a field of cooperation under an annex to the memorandum of understanding (MOU) between the Government of PNG and the Government of Australia on Cooperation in the Transport Sector. The PNG-CDP was established under this MOU in 2017, and initial activities were focused on two key areas: (i) the establishment of a strategic planning process for the PNG NWS and (ii) the provision of training to support the delivery of improved weather services and forecasting support for APEC 2018.

As part of the initial strategic evaluation, the WMO provided a meteorological observations expert to work with Bureau staff in examining the state of the observation network in more detail. The resultant report again underlined the very real challenges facing the PNG NWS but also outlined practical and useful steps forward to enhance the meteorological observation network in the country. Based on the WMO recommendations, the scope of the PNG-CDP has broadened to include provision and ongoing management of a third-party AWS observation network.

CREWS-PNG and the PNG-CDP collaborate on enhancing the meteorological observation network. Working in partnership with the PNG Remote Sensing Center (RSC), the PNG-CDP supports the PNG NWS to supplement their government-owned observation network with an existing third-party AWS network. The intention is to address the immediate need for the provision of observational data across PNG, while in parallel efforts are made to restore the existing surface-based station network to its full capacity. Over the past 10 years, the PNG RSC has created and maintained a climate monitoring network across PNG through the installation of AWSs on mobile phone towers. The network density has incrementally increased to include 37 AWSs distributed across all the climatic zones of PNG, including the majority of towns and airports. These stations have been inspected by a WMO representative to ensure that they record meteorological observations which meet the minimum threshold requirements for predictive weather and aviation purposes. A number of AWSs also support web camera technology. **Figure 8** depicts the AWS

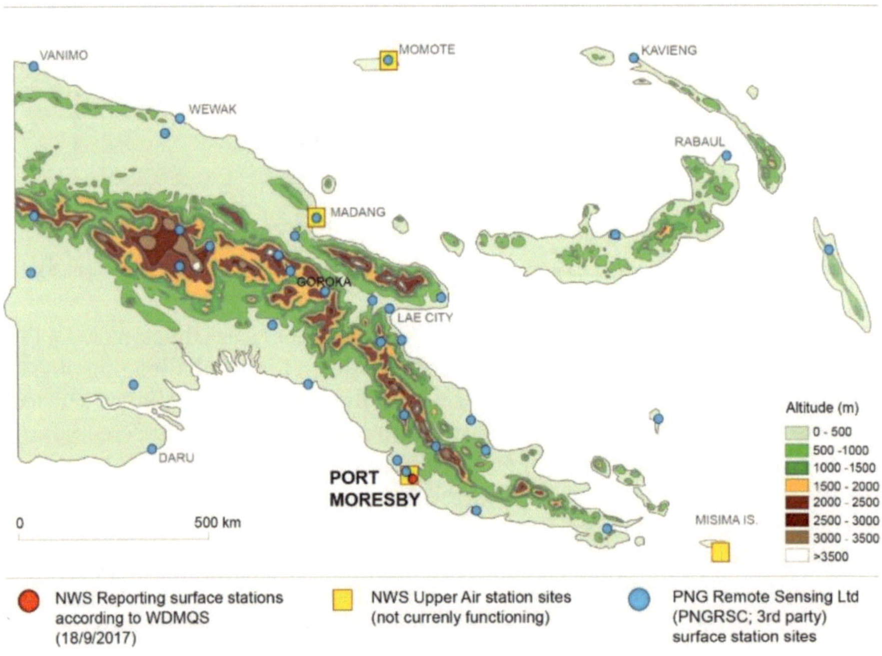

Figure 8.
Schematic map of Papua New Guinea depicting topographic extremes and the observations received from the NWS surface network on the morning of September 18, 2017, the NWS historically active upper air station sites (not currently functioning) and the location of 37 PNG Remote Sensing Centre Ltd. automated weather station sites (third party privately owned).

locations of the third-party AWS network (PNG RSC) together with the location of an existing PNG NWS surface-based stations (currently reporting) and PNG NWS upper air stations (not currently reporting).

The PNG RSC AWS data are delivered using the mobile network and a variety of web-based platforms. The PNG-CDP works with the PNG RSC to quality control the data and formats, utilize new and innovative approaches for addressing communication and Internet connectivity issues for data extraction (i.e. Amazon Web Services) and integrate the data with the PNG NWS systems to ensure it is useful, usable and used by PNG meteorologists.

A strategy for collaboration between the CREWS-PNG, the SEMDP and the PNG-CDP was discussed in detail during a workshop held in Port Moresby in February 2019. Extensive consultations with key PNG stakeholders were undertaken. Users of weather and climate information from the Department of Agriculture and Livestock, National Agriculture Research Institute, National Disaster Centre, Climate Change Development Authority, Fresh Produce Development Authority, Conservation and Environment Protection Agency, Cocoa and Coconut Research Institute, Coffee Research Institute, members of the PNG Disaster Management team and other stakeholders contributed to the development of the workshop's recommendations focusing on strengthening the capacity of the PNG NWS to provide high-quality meteorological observations, drought monitoring and prediction services addressing directly the users' needs.

Together, the CREWS-PNG, the SEMDP and the PNG-CDP, working in conjunction with the PNG NWS, will significantly strengthen capacity of the PNG NWS to provide users with valuable drought monitoring information, high-accuracy weather forecasts and subseasonal-to-seasonal climate prediction.

3. ICCAI: seasonal climate prediction for the Pacific Island nations

CREWS-PNG is building on the success of earlier Pacific projects such as the Pacific Adaptation Strategy Assistance Program (PASAP) and the Pacific-Australia Climate Change Science Adaptation Planning (PACCSAP) program undertaken under the Australian's Government International Climate Change Adaptation Initiative. Under those Pacific projects, a number of regional consolidated databases and web-based information tools were developed, to assist 15 partner countries in the Pacific with decision-making in climate change adaptation and disaster risk reduction.

It has been demonstrated that seasonal climate prediction tools are a valuable means to assist with climate change adaptation and disaster risk reduction [18]. For decades, seasonal climate prediction in Australia and the Pacific Island Countries was based on statistical models [10]. However, in a rapidly changing environment, the skill of statistical models which are solely based on past records is expected to deteriorate. Hence dynamical climate models, which are based on laws of physics rather than past relationships, are preferable for forecasting climate on time scales from weeks to months and beyond.

It has been shown that using the outputs of the dynamical climate model Predictive Ocean and Atmosphere Model for Australia (POAMA), skilful predictions of regional rainfall could be provided 2–3 months in advance [19, 20]. POAMA was developed [21] and used by the Bureau of Meteorology operationally to produce its seasonal climate outlooks from 2013. It is a state-of-the-art dynamical (physics-based) forecast model which used ocean, atmosphere, ice and land observations to initiate the model and produce outlooks for the season ahead.

Recently, a new higher resolution model—the Australian Community Climate and Earth-System Simulator (ACCESS)—has succeeded POAMA to further enhance the Bureau's capability in weather and climate prediction. Since August 2018 the Bureau's operational climate forecast system for multi-week, monthly, seasonal and longer-range climate outlooks is ACCESS-Seasonal (ACCESS-S). The atmosphere and land model components of ACCESS-S operate at an approximate global resolution of 60 km, providing far greater detail than POAMA which had an approximate resolution of 250 km. The multi-week and seasonal performance of ACCESS-S has been evaluated for Australia based on a 23-year hindcast set and compared to the previous operational system, POAMA [22]. However, hindcast skill and calibration have not been assessed for other countries in the region including PNG. One of the key tasks for CREWS-PNG is to assess and utilize outputs of ACCESS-S in order to produce accurate multi-week and seasonal climate outlooks that will improve drought prediction for the country.

4. Web-based information tools to disseminate climate monitoring and seasonal prediction products

CREWS-PNG is a successor of the Pacific projects of the ICCAI, and its implementation strategy is based on similar approaches to develop climate monitoring and seasonal prediction products as well as means of disseminating information. It is important to ensure that climate monitoring and subseasonal-to-seasonal prediction products are not only of high accuracy but also are delivered to users in a timely manner and in a way that can be utilized in their decision-making. As part of the PASAP and PACCSAP activities, the Pacific Seasonal Prediction Portal was developed to disseminate climate data and dynamical climate model-based forecast products through a range of web-based information tools (**Figure 9**). This portal could be accessed through the Australian Bureau of Meteorology's website: http://www.bom.gov.au/climate/pacific/projects.shtml.

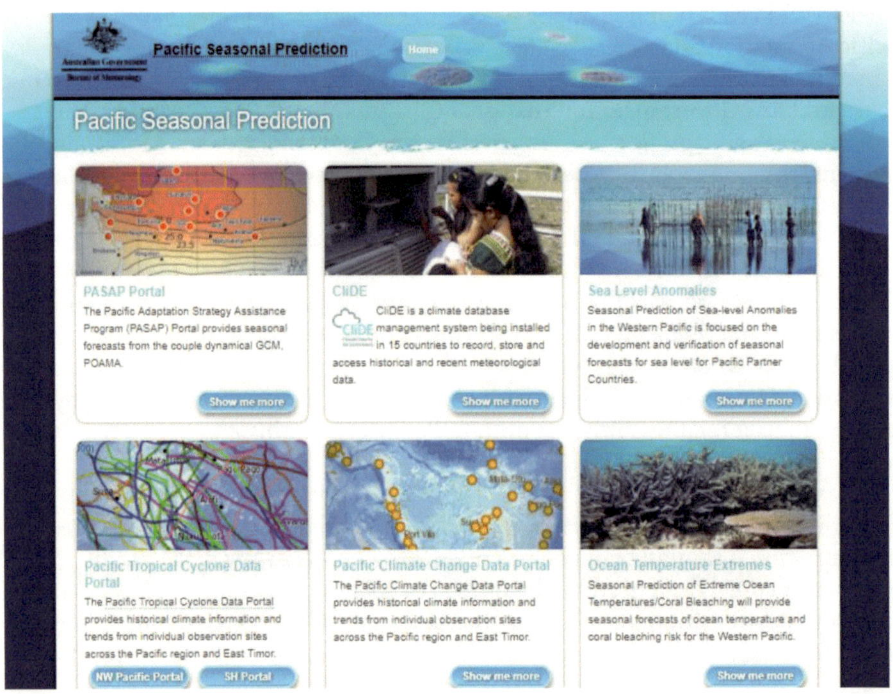

Figure 9.
Home page of the Pacific Seasonal Prediction Portal.

The Pacific Seasonal Prediction Portal provides users with access to a range of products for monitoring and prediction of weather and climate extremes. Here we briefly introduce some of these products, including key features of the tropical cyclone (TC) and seasonal prediction portals, including extreme ocean temperatures and potential coral bleaching.

TCs are severe weather events which affects Pacific Island Countries every year, causing loss of life land property. Damage caused by TCs is not only caused by destructive winds but also torrential rain, high ocean waves and storm surge. To be prepared for the multi-hazards associated with TCs, knowledge about regional historical cyclone tracks are important. Meteorological agencies of the Pacific Island Countries as well as numerous other users utilize the portal to access such information for areas of interest. As an example, tracks of 80 tropical cyclones which passed through exclusive economic zone of PNG between the 1969–1970 and 2016–2017 tropical cyclone seasons retrieved through the Southern Hemisphere Tropical Cyclone Data Portal are presented in **Figure 10**. A comprehensive description of the Southern Hemisphere Tropical Cyclone Data Portal could be found in [23], which also provides readers with the users' guide to the web-based information tool to select, retrieve, display and analyse TC data.

The importance of seasonal climate prediction to assist with decision-making is recognized by users from climate-sensitive sectors around the world. Seasonal climate outlooks at various levels—national, regional and global—are operationally produced by NMHSs, Regional Climate Centres (RCCs) and the WMO providing users with vital information about state of the El Niño-Southern Oscillation and its expected development, plus seasonal climate outlooks for temperatures, rainfall and a variety of other variables. Climate outlooks at the global scale are disseminated by the WMO, while at a regional level, WMO GPC LRFs and RCCs play a leading role in this task. In the Pacific, WMO GPC LRF Melbourne is tasked with disseminating outputs from the dynamical climate model POAMA to RCCs and NMHSs in the region. As an example, the seasonal prediction of sea surface temperature in the Pacific for January to March 2019 is presented in **Figure 11**, demonstrating a significant oceanic warming in equatorial central Pacific—a possible precursor to El Niño.

The livelihood of many people in Pacific Island Countries is highly dependent on the productivity of the oceans including coral reefs which surrounds the islands. A

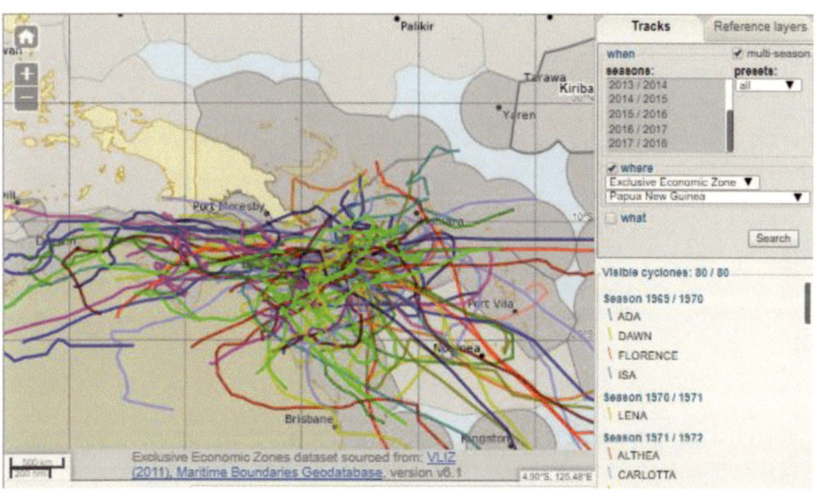

Figure 10.
Tracks of 80 tropical cyclones from the Southern Hemisphere Tropical Cyclone Data Portal which passed through the exclusive economic zone (EEZ) of PNG between the 1969–1970 and 2016–2017 tropical cyclone seasons.

dramatic impact of climate change on the health of coral reefs has been observed in recent decades due to the increase in ocean acidification and especially due to increase in frequency and intensity of extreme ocean temperatures which led to severe coral bleaching events and impacts on marine life [24]. Disseminating early warnings

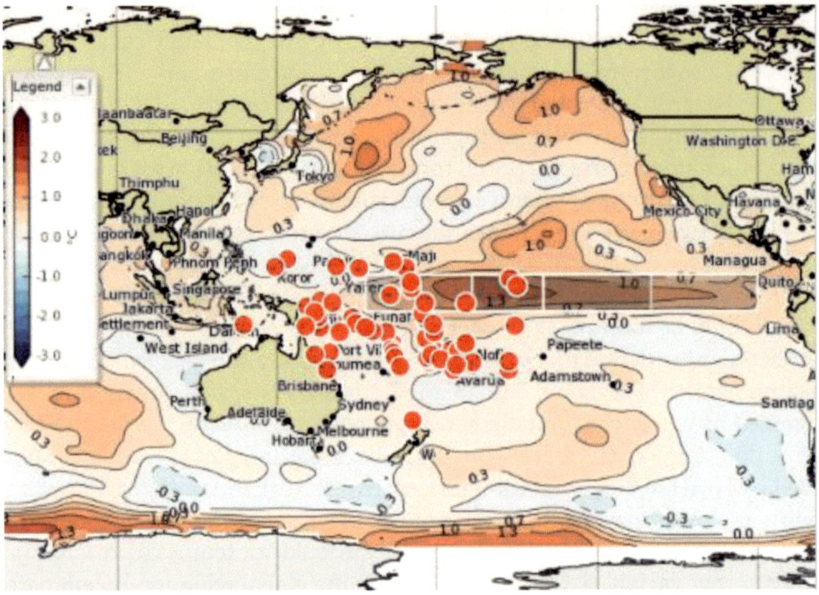

Figure 11.
Seasonal prediction of sea surface temperature in the Pacific for January–March 2019 based on outputs of POAMA retrieved through the Pacific Seasonal Prediction Portal—WMO GPC LRF Pacific Seasonal Prediction.

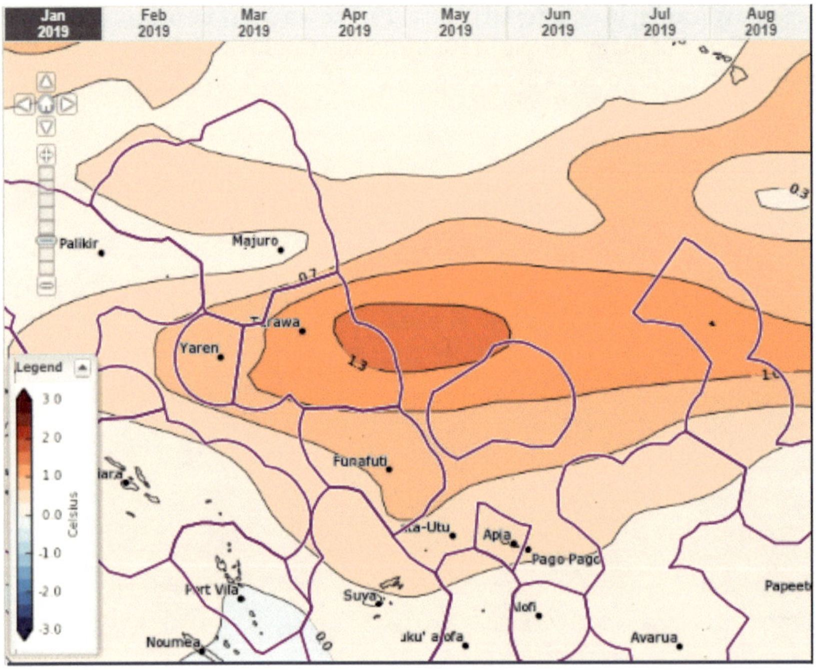

Figure 12.
Seasonal prediction of sea surface temperatures around Kiribati based on outputs of POAMA retrieved through the Pacific Seasonal Prediction Portal—extreme ocean temperatures and coral bleaching.

about possible extreme ocean temperatures, e.g. because of a coming El Niño event, is essential—it provides local government authorities with vital information 2–3 months in advance allowing them to implement protective measures. For example, the spatial distribution of predicted sea surface temperatures around Kiribati is presented in **Figure 12** (the forecast for January 2019 is based on outputs of POAMA; model was initialized on December 02, 2018). Extreme ocean temperatures in some areas of exclusive economic zone of Kiribati are expected to exceed 1.3°C compared to climatology (1982–2010), potentially leading to coral bleaching in those areas.

CREWS-PNG is taking the outlined above approach to disseminate climate information based on outputs of the new ACCESS-S dynamical climate model as part of its implementation strategy. The introduction of impact-based drought forecasts and associated risk-based warnings for improved decision-making is planned to be delivered through a specialized web-based information tool which will be an integral part of the Pacific Seasonal Prediction Portal. Products and climate services will be developed and implemented based on close consultation with stakeholders to understand their decision-making requirements.

5. Conclusions

Climate monitoring combined with skilful subseasonal-to-seasonal climate prediction enables NMHS to provide their governments, industry sectors and local communities with accurate information to assist in their decisions around how to adapt to climate variability and change. CREWS-PNG will contribute significantly to enhance EWS in PNG, and it is envisaged that cutting edge weather and climate monitoring and prediction products developed for PNG under this project could be expanded to include NMHS across the Asia-Pacific region. The result would be a significant advance for climate services provided by meteorological agencies of LDCs and SIDS in the region.

Acknowledgements

Climate Risk and Early Warning Systems international initiative and the WMO provided financial support for the CREWS-PNG project. The Australian Government Department of Foreign Affairs and Trade (DFAT) funded the PNG-CDP project. Projects of the Pacific Adaptation Strategy Assistance Program and the Pacific-Australia Climate Change Science Adaptation Planning program of the International Climate Change Adaptation Initiative were financially supported by the Australian Agency for International Development (AusAID), the Australian Government Department of Climate Change and Energy Efficiency (DCCEE) and the DFAT.

Conflict of interest

Declaration: there is no conflict of interest.

Author details

Yuriy Kuleshov[1,2,3*], Kasis Inape[4], Andrew B. Watkins[1], Adele Bear-Crozier[1], Zhi-Weng Chua[1], Pingping Xie[5], Takuji Kubota[6], Tomoko Tashima[6], Robert Stefanski[7] and Toshiyuki Kurino[7]

1 The Australian Bureau of Meteorology, Melbourne, Australia

2 Royal Melbourne Institute of Technology (RMIT) University, Melbourne, Australia

3 The University of Melbourne, Melbourne, Australia

4 Papua New Guinea National Weather Service (PNG NWS), Port Moresby, PNG

5 The Climate Prediction Center, National Oceanic and Atmospheric Administration (NOAA), Washington DC, USA

6 The Earth Observation Research Center, Japan Aerospace Exploration Agency (JAXA), Tsukuba, Japan

7 The World Meteorological Organization, Geneva, Switzerland

*Address all correspondence to: yuriy.kuleshov@bom.gov.au

IntechOpen

© 2019 The Author(s). Licensee IntechOpen. This chapter is distributed under the terms of the Creative Commons Attribution License (http://creativecommons.org/licenses/by/3.0), which permits unrestricted use, distribution, and reproduction in any medium, provided the original work is properly cited.

References

[1] Stocker TF, Qin D, Plattner G-K, Tignor M, Allen SK, Boschung J, et al., editors. IPCC, 2013: Climate Change 2013: The Physical Science Basis. Contribution of Working Group I to the Fifth Assessment Report of the Intergovernmental Panel on Climate Change. Cambridge, United Kingdom/New York, NY, USA: Cambridge University Press; 2013. DOI: 10.1017/CBO9781107415324. 1535pp

[2] Asia-Pacific Disaster Report 2012—Reducing Vulnerability and Exposure to Disasters. UNESCAP [Internet]. 2012. Available from: https://www.unescap.org/publications/asia-pacific-disaster-report-2012-reducing-vulnerability-and-exposure-disasters [Accessed: November 30, 2018]

[3] Asia-Pacific Disaster Report 2017—Live No One Behind. UNESCAP [Internet]. 2017. Available from: https://www.unescap.org/publications/asia-pacific-disaster-report-2017-leave-no-one-behind [Accessed: January 13, 2019]

[4] Disasters Could Cost Asia-Pacific Region $160 Billion per year by 2030, UN Warns [Internet]. 2018. Available from: https://news.un.org/en/story/2018/04/1008182 [Accessed: January 13, 2019]

[5] Kuleshov Y, McGree S, Jones D, Charles A, Cottrill A, Prakash B, et al. Extreme weather and climate events and their impacts on island countries in the Western Pacific: Cyclones, floods and droughts. Atmospheric and Climate Sciences. 2014;**4**:803-818. DOI: 10.4236/acs.2014.45071

[6] Information Bulletin—Tuvalu Drought [Internet]. 2011. Available from: http://www.ifrc.org/docs/appeals/rpts11/IBTV14101102.pdf [Assessed: January 13, 2019]

[7] Papua New Guinea El Nino 2015-16 [Internet]. 2015. Available from: https://reliefweb.int/sites/reliefweb.int/files/resources/PNG%20Brief%20Sep2015.pdf [Assessed: January 13, 2019]

[8] CREWS [Internet]. 2015. Available from: http://www.ifrc.org/docs/appeals/rpts11/IBTV14101102.pdf [Assessed: January 13, 2019]

[9] CREWS Expands [Internet]. 2018. Available from: https://public.wmo.int/en/media/news/climate-risk-and-early-warning-systems-initiative-expands-its-coverage [Assessed: January 13, 2019]

[10] Cottrill A, Kuleshov Y. An assessment of rainfall seasonal forecasting skill from the statistical model SCOPIC using four predictors. Australian Meteorological and Oceanographic Journal. 2014;**64**:273-281

[11] Fawcett RJB, Jones DA, Beard GS. A verification of publicly issued forecasts issued by the Australian Bureau of Meteorology: 1998-2003. Australian Meteorological Magazine. 2005;**54**:1-13

[12] Kuleshov Y, Kurino T, Kubota T, Tashima T, Xie P. WMO Space-based Weather and Climate Extremes Monitoring Demonstration Project: Early Outcomes of Asia-Pacific Regional Cooperation on Drought and Extreme Precipitation Monitoring from Space". In: Rainfall. London, United Kingdom: IntechOpen; 2019. (in press)

[13] Xie P, Joyce R, Wu S, Yoo S-H, Yarosh Y, Sun F, et al. Reprocessed, bias-corrected CMORPH global high-resolution precipitation estimates from 1998. Journal of Hydrometeorology. 2017;**18**(6): 1617-1641. DOI: 10.1175/JHM-D-16-0168.1

[14] Kubota T, Shige S, Hashizume H, Aonashi K, Takahashi N, Seto S, et al. Global precipitation map using satellite-borne microwave radiometers by the GSMaP project: Production and validation. IEEE Transactions

on Geoscience and Remote Sensing. 2007;**45**(7):2259-2275

[15] McKee T, Doesken NJ, Kleist J. The relationship of drought frequency and duration of time scales. In: Eighth Conference on Applied Climatology; 17-23 January 1993; Anaheim, CA: American Meteorological Society; 1993. pp. 179-186

[16] Kogan FN. Operational space technology for global vegetation assessment. Bulletin of the American Meteorological Society. 2001;**82**(9):1949-1964. DOI: 10.1175/1520-0477(2001)082<1949:OSTFGV>2.3.CO:2

[17] Cold and Warm Episodes by Season. [Internet]. 2018. Available from: https://origin.cpc.ncep.noaa.gov/products/analysis_monitoring/ensostuff/ONI_v5.php [Assessed: January 13, 2019]

[18] Charles A, Kuleshov Y, Jones D. Managing climate risk with seasonal forecasts. In: Banaitiene N, editor. Risk Management—Current Issues and Challenges. Rijeka, Croatia: InTech; 2012. pp. 557-584. DOI: 10.5772/2568

[19] Cottrill A, Hendon H, Lim E-P, Langford S, Shelton K, Charles A, et al. Seasonal forecasting in the Pacific using the coupled model POAMA-2. Weather and Forecasting. 2013;**28**:668-680. DOI: 10.1175/WAF-D-12-00072.1

[20] Charles AN, Brown JR, Cottrill A, Shelton KL, Nakaegawa T, Kuleshov Y. Seasonal prediction of the South Pacific convergence zone in the austral wet season. JGR-Atmospheres. 2014;**119**(22):12,546-12,557. DOI: 10.1002/2014JD021756

[21] Hendon HH, Lim E-P, Wang G, Alves O, Hudson D. Prospects for predicting two flavours of El Nino. Geophysical Research Letters. 2009;**36**(19):L19713. DOI: 10.1029/2009GLO40100

[22] Hudson D, Alves O, Hendon HH, Lim E, Liu G, Luo JJ, et al. ACCESS-S1: The new Bureau of Meteorology multi-week to seasonal prediction system. Journal of Southern Hemisphere Earth Systems Science. 2017;**67**(3):132-159. DOI: 10.22499/3.6703.001

[23] Kuleshov Y. "Climate change and southern hemisphere tropical cyclones" International initiative: Twenty years of successful regional cooperation. In: Handbook on Weather and Climate Extremes. Berlin, Germany: Springer; 2019 (in press)

[24] Kuleshov Y, Jones D, Hendon H, Charles A, Cottril A, Lim EP, et al. Pacific Adaptation Strategy Assistance Program: Strengthening the capacity for seasonal prediction services in Pacific countries. Bulletin of the Australian Meteorological and Oceanographic Society. 2012;**25**(1):7-12

Chapter 10

Benefits of Arbuscular Mycorrhizal Fungi Application to Crop Production under Water Scarcity

Katalin Posta and Nguyen Hong Duc

Abstract

Water deficit is one of the most severe abiotic stresses threatening crop growth and production on the globe. Water stress causes a series of morphological, biochemical, physiological, and molecular alterations that negatively influence plant productivity. However, in nature, plants are often associated with microbes that can modulate plant responses to water scarcity. Among beneficial microbes, arbuscular mycorrhizal fungi (AMF) are one of the most widespread symbiotic fungi colonizing the majority of agricultural plants. Besides an enhancement in plant nutrition, AMF have been reported to improve plant performance under water restrictions. In this chapter, we emphasize the benefits of AMF inoculation to crop production under water deficit based on related laboratory and field experiments. Variable outcomes and challenges of AMF application are also discussed for practical use in crop production under water scarcity.

Keywords: arbuscular mycorrhizal fungi, drought stress, water relations, crop productivity, plant tolerance, drought mitigation

1. Introduction

Water scarcity jeopardizes not only originally arid, semi-arid regions but also agricultural areas in which farmers obtain flourishing horticulture based on adequate water resources. Nonetheless, ongoing climate change supposed to amplify the frequency and severity of drought in different regions of the globe [1] can wipe out the so far achievements. Drought is the most devastating stress that remarkably diminishes crop productivity more than any other stress factor [2]. Water constraints provoke stomatal closure with a subsequent reduction of CO_2 influx resulting in a decrease in photosynthetic activity and carbon partitioning [3]. Also, water scarcity has a negative influence on nutrient supply, reducing phosphate availability. Severe drought profoundly affects plant physiology, growth, development, and reproduction, and exerts substantial losses in crop yield as well as reduces crop quality. In fact, over the past 35 years, worldwide drought inflicted yield decrease by 40% in maize and 21% in wheat production [4]. Thus, there is an urgent need to develop strategies to make agriculture more resilient and to alleviate the adverse impacts of water scarcity on crop yield. Among these strategies, there has been an increasing interest in beneficial soil microbes including arbuscular mycorrhizal (AM) fungi.

IntechOpen

Notably, under natural conditions, plants frequently interact with microbes, which directly mediate plant responses to environmental adversities. Some microbe-plant interactions lead to a mitigation of stress-related damages and improvement of plant tolerance to stressful conditions [5]. As a crucial element of soils, microbes are an integral part of the agricultural ecosystem. Arbuscular mycorrhizal fungi (AMF) are ubiquitous soil microorganisms, which can form a symbiotic association with most terrestrial plants. These beneficial microbes have been proved to offer an array of benefits to host plants [6]. During mycorrhization, besides significant improvement of plant nutritional status, AMF can enhance plant performance and tolerance against several stresses, particularly drought stress [7]. The exploitation of AMF is considered as one of the most efficient practices to increase plant tolerance to environmental stresses [8]. Previous studies illustrate the substantial contribution of AM symbiosis to improved stress plant tolerance to water deficit by various mycorrhizal benefits such as strengthened water and nutrient uptake, alterations in host physiology, for example, photosynthesis, osmotic adjustment, phytohormones, and more efficient antioxidative systems [9–11]. This chapter presents the current knowledge on AMF application to crop production under water deficit. Variable benefits of AMF are also discussed to explain the reason why positive outcomes of AM colonization are not always the case. Finally, challenges of the fungal symbiont application are highlighted for practical use in crop production.

2. General features of AMF

AMF are obligate root symbionts inhabiting almost all terrestrial ecosystems. They can form a symbiotic association with around 80% of vascular plants and with approximately 90% of agricultural plants [12]. In this mutual association, the fungus receives 10–20% of total photosynthates [13] and lipids [14] from the host plant, whereas the plant is enhanced through uptake of water and mineral nutrient by the mycorrhizal partner [12]. AMF are the most common fungi in soils and represent 9–55% of the soil microbe biomass and 5–36% of the total soil biomass [15]. These fungi play a vital role in agricultural ecosystems, since they can improve plant nutrient, water status, and plant growth [12], enhance survival rate and development of seedlings, crop uniformity, and reproductive capacity [16], decrease the input of P and N fertilizer, and increase resistance or tolerance to environmental adversities [8, 17].

Currently, AMF are classified as a member of phylum Glomeromycota including four orders (Archaeosporales, Diversisporales, Glomerales, and Paraglomerales), with 11 families, 25 genera, and nearly 250 species [18]. However, data based on next-generation sequencing of root samples [19] and recent results [20] suggest that its number may be an order of magnitude higher. Spores of AMF which are the major survival units of AMF have multi-nucleate, heterokaryotic structures [21], and are formed singly, in clusters or sporocarps in the soil, and within root tissue in some mycorrhiza species as well (**Figure 1A–C**). The development of AM symbiosis starts with signaling taking place before physical contact between the plant and the fungus. Both partners produce molecular signals triggering preparative responses in the other [22]. The mycorrhization process can be divided into distinct steps, consisting of germinating spores, hyphae differentiation, appressorium formation, penetration of the host root, intraradical hyphae formation, intercellular growth along with developed external mycelium (extraradical hyphae), and arbuscule formation, subsequently exchanging nutrients and carbohydrates between the host and fungus [23].

The primary structures of AMF consist of coenocytic hyphae with unlimited growth in the rhizosphere called external hyphae, which penetrate the cortex

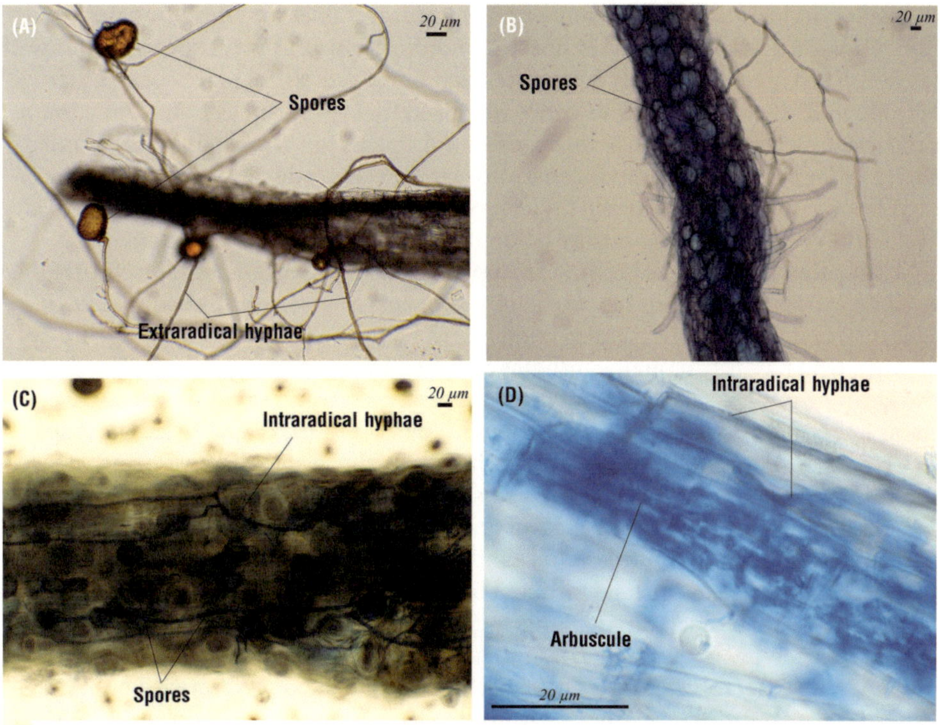

Figure 1.
Tomato roots without (A) and with (B–D) staining showing AM fungal structures. The presence of arbuscular mycorrhiza (AM) structures (arbuscules, vesicles, intraradical hyphae, and spore) was assessed by means of an Olympus BX51 light microscope with Nomarski interference contrast optics, using an objective of 40×. Scale Bar representing 20 μm.

layer of roots and form different organs. The extraradical hyphae merely in some species of Diversisporales [18] producing auxiliary cells could have functions in reproduction or nutrition and storage [24]. Mycelium outside the roots absorb mineral nutrients and water and subsequently transport them to the host plant via intraradical hyphae (**Figure 1C, D**) growing inside root cells [6]. Hyphae growing within roots form either the Paris-type or the Arum-type. The Paris-type is featured by intracellular mycelium development to shape coils, whereas the Arum-type is characterized by intercellular hyphae growth forming arbuscules [12] (**Figure 1D**), thereby establishing the nutrient exchange sites between AMF and the host plant [25]. Vesicles containing high quantity of lipids and glycogen are formed from intraradical hyphae at intercalary position (their terminal) in the root, functioning as nutrient storage, and propagules [23] but not all AMF produce vesicles.

AMF species isolates differ in the ability to spread mycelia, the viability, structure, and possibility of anastomosis [26, 27]. Taxonomic variation in mycelium structure among AMF families was also observed [28]. *Gigasporaceae* are prone to possess vigorous, thickly aggregated mycelium with densities from 6 to 9 m cm^{-3}, while *Acaulosporaceae* and *Glomeraceae* show a tendency to maintain thinly dispersed mycelium with densities from 1 to 2 m cm^{-3}.

3. Variable crop responses to AMF

Although a majority of plants are responsive to AMF, plant species in families *Amaranthaceae, Brassicaceae, Caryophyllaceae, Chenopodiaceae, Cyperaceae,*

Juncaceae, and *Urticaceae* are rarely or never colonized by the symbiotic fungus [29]. How AMF evaluate the AM host and nonhost status of plant species is not well known. The current hypothesis proposes that nonmycorrhizal plant species lost orthologs of important putative genes, required for symbioses [30], and/or cannot synthesize or degrade strigolactones, essential signals for symbiosis establishment [31], and/or their root exudates constitute antifungal products [29]. Under certain conditions, some nonhost species develop rudimentary AM phenotypes described by Cosme et al. [30] giving a more in-depth explanation of this question.

Utilization of AMF has become an appealing tool for sustainable agriculture due to the positive attributes of mycorrhizal symbiosis. Nevertheless, the opposite or neutral influence of AMF has also been found [32]. The obligate biotrophic life cycle of AMF which relies on photosynthates supplied by a nurturing autotrophic host is the key point; therefore, choosing the right partner (target plant) is crucial. Even though this widespread symbiont is thought to be a generalist due to low host specificity, each AMF species highly varies in the responsiveness to the host plant. Hence, the variable benefits of AM symbiosis exist among mycorrhiza species [10, 33]. The interaction between the host plant and AMF could range from mutualism to parasitism in which colonized plants exhibit a decrease in growth [34] owing to the carbon drainage in the host inflicted by the fungus [35]. Many factors that can affect the AM benefits to target plants include host plant genotypes, AMF species, and environmental conditions. Dissimilar plant responses to different AMF species under environmental adversities have been observed [11, 36]. Fascinatingly, AM benefits for plant fitness augment with adversity, supporting the concept of AM colonization as a 'health insurance' for host plants, in which the beneficial effects of AMF become more obvious under stressful environments [36]. Metabolites differentially accumulated in roots colonized by different fungal symbionts (*Rhizophagus irregularis*, *Funneliformis mosseae*, and *Claroideoglomus etunicatum*) under abiotic stresses, which may underlie their enhanced stress tolerance in host plants [36]. Cultivar differences in response to mycorrhizas have been reported in many crops such as tomato [37], pepper [38], wheat [39], maize [40], and some other crops [41]. For chickpea, only three of thirteen varieties with different genotypes and phenotypes were more positively responsive to AM mixed inoculation with *Diversispora eburnea*, *Claroideoglomus etunicatum*, and *Glomus* sp. [42]. More recently, twenty geographically different barrel clover (*Medicago truncatula*) accessions showed differences in their growth, stomatal conductance (g_s), and AM colonization in response to *Funneliformis mosseae* treatment [43]. Also, root hydraulic conductivity, expression of the mycorrhiza-induced phosphate transporter gene (*MtPT4*), and five aquaporin genes (*MtAQP1, MtPIP1, MtPIP2, MtNIP1,* and *MtNIP4*) vary with mycorrhizal treatment during further analysis of five accessions. In the case of wheat, old accessions have been shown to be more responsive to AMF than new ones [39].

Selection and breeding programs generally tend to maximize plant performance and crop yield under high-input production systems, which could cause the loss of genes, phytochemicals, and/or other plant traits which are necessary for the establishment of efficient symbioses. Modern cultivars could absorb phosphate without the AM assistance in soils with high phosphorus availability, decreasing the degree of AM dependence. As a consequence, AMF are less responsive to new lines. Recent research has proved that domestication decreased AM benefits for domesticated crops in exposure to high P supply [44]. However, in maize, which is highly mycorrhizal-dependent, modern breeding programs do not necessarily result in the less mycorrhizal colonization. Replicated field experiments with 225 genotypes consisting of hybrids, inbred lines, and landraces originating from different locations were conducted for two consecutive years to explore the variation in mycorrhizal colonization [40]. The findings showed that AM colonization differed profoundly and continuously among genotypes, with substantially greater values in modern hybrids than old landraces and inbred lines.

4. Mechanisms of AMF mitigate drought stress in host plants

It is well known that AMF offer indispensable advantages to the host plant subjected to water shortage, with two major strategies that mycorrhizal plants use to deal with water deficit: drought mitigation and drought tolerance. Drought mitigation strategy is involved in indirect AM benefits and enhanced water uptake through the extensive hyphae network, enabling host plants to suffer less stress than non-AM plants, whereas drought tolerance includes a combination of direct AM benefits that improve plant's innate ability to cope with the stress (**Figure 2**).

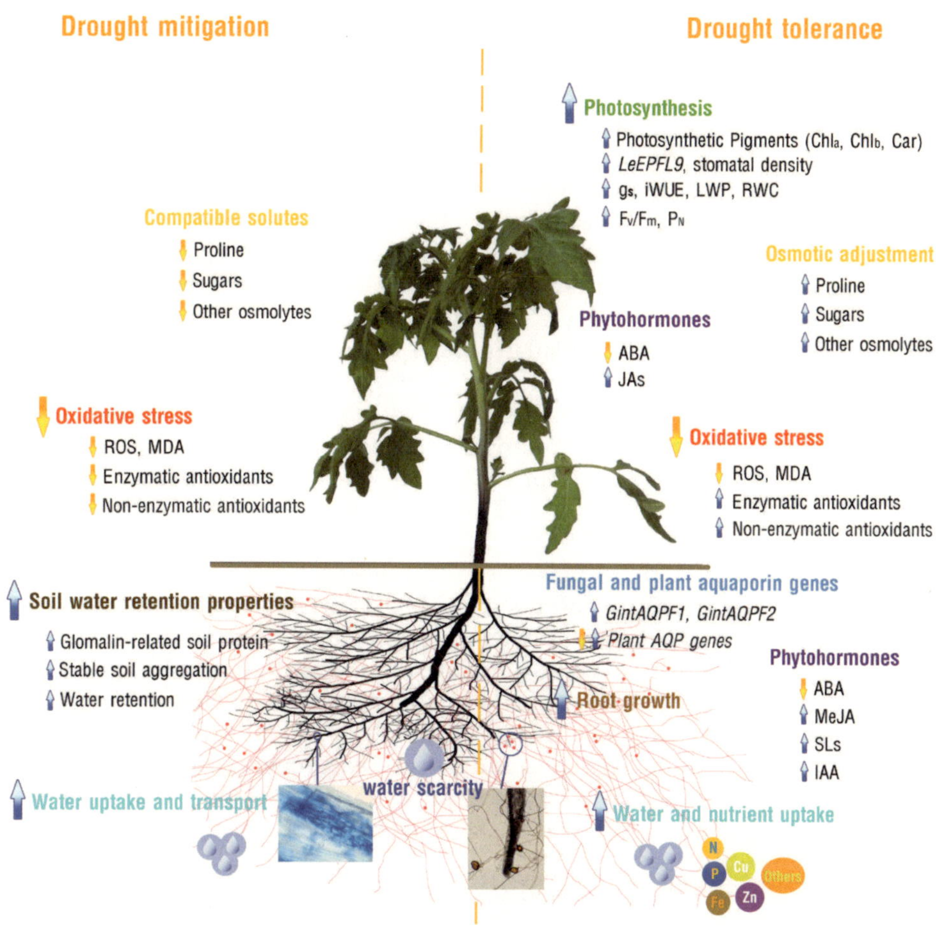

Figure 2.
Strategies of mycorrhizal plants to cope with water scarcity, that is, drought mitigation and drought tolerance. Multiple benefits/mechanisms could be simultaneously induced by arbuscular mycorrhizal fungi in the host plant exposed to water deficit. The blue arrows show increase/up-regulation, whereas the orange arrows indicate decrease/down-regulation, relative to control non-mycorrhizal plants. Italic words indicate genes. ABA, abscisic acid; AQP, aquaporin; Car, carotenoids; Chl_a, chlorophyll a; Chl_b, chlorophyll b; F_v/F_m, maximum quantum efficiency of PSII; g_s, stomatal conductance; IAA, indole-3-acetic acid; iWUE, intrinsic water use efficiency; JAs, jasmonates; LWP, leaf water potential; MDA, malondialdehyde; MeJA, methyl jasmonate; P_N, net photosynthesis rate; ROS, reactive oxygen species; RWC, relative water content; SLs, strigolactones.

4.1 Direct benefits of AM symbiosis for host plants under water deficit

4.1.1 Improved water and nutrient uptake through the hyphal network of AMF

An important benefit of AM colonization to the host plant under drought stress is a superior water allocation mediated by the fungal hyphal network, facilitating the colonized root access to water in a lower soil water potential [45]. Indeed, the host root system is extended by widespread extraradical mycelia, enabling colonized roots to reach more water and nutrient pools unavailable to uncolonized roots. Fungal hyphae diameters (3–7 μm) are much smaller than those of fine root hairs (5–20 μm); nevertheless, hyphal densities are ten-hundred times higher than root densities [46]. Hence, the absorption surface of mycorrhizal roots is improved substantially. It is calculated that the rate of water transport from external hyphae to the root ranged from 0.1 [47] to 0.76 μl H_2O h^{-1} per hyphal infection point [48], which is adequate to alter plant water relations [47]. Lettuce plants pretreated by *Rhizophagus irregularis*, *Funneliformis mosseae*, *Funneliformis coronatum* (formerly *Glomus coronatum*), and *Claroideoglomus claroideum* (*G. claroideum*) obtained 3–4.75 ml H_2O $plant^{-1}$ day^{-1} higher than uncolonized plants, which might be related to the amount of extraradical mycelium and root colonization frequency [45]. Furthermore, AMF contribute approximately 20% to total plant water uptake [49], highlighting the role of the symbiosis in the water status of host plants.

The widespread extraradical mycelia also enhance the absorption of mineral nutrients in soils, which is more critical for host plants under water-stress conditions where nutrient mobility is limited. As soon as external hyphae transport water to the host, mineral nutrients also follow the water flow to the plant from the soil-root interface [50]. AM colonization is well known to improve phosphorus (P) nutrient into the host plants particularly under low-nutrient conditions, increasing stress tolerance in plants. Interestingly, plants possess a symbiotic inorganic phosphate (Pi) uptake pathway, and AM symbiosis has been proved to specifically induce the expression of genes encoding plant Pi transporters to enhance P acquisition, for instance, *LjPT4* in *Lotus japonicus* and *MtPT4* in *Medicago truncatula* [51], recently *LbPT3, LbPT4,* and *LbPT5* in *Lycium barbarum* [52]. Under water restrictions (moderate and severe), different expressions of five tomato *PT* genes (*LePT1-LePT5*) in the absence/presence of *Rhizophagus irregularis* or *F. mosseae* were observed [53]. *LePT4* was overexpressed in *R. irregularis*-colonized plants exposed to both water-stress levels, while this upregulation was in *F. mosseae*-infected plants subjected to severe water stress. A role of *PT4* genes in root tips, creating a connection among root branching, Pi-signaling mechanisms, and Pi-perception has been proposed [51]. In addition, on the fungal side, *R. irregularis PT* gene was up-regulated under moderate drought conditions [53]. Phosphate is taken up by mycorrhizal phosphate transporters and assimilated to polyphosphate translocated toward the plant. This process is facilitated by the activation of fungal aquaporins [54].

Apart from that, AM colonization enhances the rate of nitrogen (N)-assimilation of plants under drought [55] as a result of the direct uptake of NO_3^- or NH_4^+ by fungal hyphae [56]. Several NO_3^- and NH_4^+ transporters and metal transporters in AMF [57, 58] while mycorrhiza-inducible ammonium transporters in some plants have been identified [59, 60]; therefore, AMF considerably contribute to the total N uptake of the host. Increased N nutrient could promote protein synthesis and higher levels of compatible osmolytes in stressed AM plants. Other studies also confirmed that inadequacy of necessary macro- and micro-nutrients could be alleviated in mycorrhizal plants under water deficit [61, 62]. Hydraulic conductivity of colonized roots was enhanced to absorb more N, P, and K, leading to a higher protein concentration in host plants under drought stress [63]. Thus, more vigorous uptake of water and nutrients may provide adequate necessary substances for better growth of mycorrhizal plants under such stress.

4.1.2 AMF-induced changes in expression of aquaporin genes, transcriptional profiles

The negative water potential in dried soils exerts the problem for plants to obtain adequate water amount, a process where aquaporins (AQPs) get involved in [64]. AQPs belonging to the large major intrinsic protein family of transmembrane proteins functioning as water channels are crucial in osmoregulation [64]. On top of that, their regulation of transcellular movement of many molecules such as small alcohols, boron, and osmolytes has been reported [65]. In AMF, the first AQP gene *GintAQP1* of *Rhizophagus irregularis* was cloned, with evidence of a compensatory mechanism between *GintAQP1* expressions and the host aquaporins under drought stress [66]. Furthermore, two AQP genes *GintAQPF1* and *GintAQPF2* present in *Rhizophagus irregularis* were upregulated under osmotic stress, assisting the fungus survival and contributing to the host plant tolerance to water stress [67, 68]. Upregulation of *RiAQPF2* in *Rhizophagus irregularis* was also found under water deficit [10], suggesting its putative involvement in host plant tolerance in response to drought.

On the plant side, AMF could induce changes in the expression of various AQP genes in the host in order to strengthen root hydraulic conductivity and host tolerance under water-stress conditions in several plants, such as maize [69–71], tomato [10, 11], black locust [72], trifoliate orange [73], olive [74], and *Populus x canadensis* plants [75]. AM-induced alterations in expression of plant AQPs could depend on stress duration as the observation in maize plants [69]. Under short-term water deficit, the AM symbiosis upregulated ten AQP genes with diverse aquaporin classes in roots inoculated with *Rhizophagus intraradices*, stimulating more water uptake in the host [69]. By contrast, under sustained water-stress conditions, AM-mediated downregulation of 6 different AQP genes was found, restricting plant water loss [69]. Intriguingly, drought-sensitive cultivars may gain higher physiological benefit from AM inoculation than drought-tolerant cultivars [71]. Downregulation of genes *TIP1;1, TIP2;3, PIP1;1, PIP1;3, PIP1;4, PIP1;6, PIP2;2*, and *PIP2;4* whereas only upregulation of *TIP4;1* were observed in drought-sensitive cultivar colonized by *Rhizophagus irregularis*, supporting the decrease in water loss in host plants subjected to drought stress [71]. Recent research also revealed a significant shift in the transcriptional regulation profiles with AQP genes as potential targets in mycorrhizal roots, in comparison to non-AM ones during a water stress event, which may influence some key metabolic pathways linked with drought response [76]. In parallel, it has been proposed that during drought stress a controlled mechanism mediated by the presence of arbuscules at cortical cells in roots fine-tuned the gene expression regulation in the host plant [76].

In general, fungal and plant AQPs work together in mycorrhizal plants under water restrictions. The simultaneous induction of both fungal and plant AQP genes together with differential regulation of drought-responsive genes in host plant indicates that AMF mediate colonized plant responses to drought stress.

4.1.3 Increased photosynthetic efficiency

Numerous reports illustrate that AMF could increase photosynthetic activity or protect the photosynthetic apparatus under water stress conditions [77, 78]. In fact, AM colonization considerably influences the stomatal behavior in the leaves of host plants, determining the water vapor efflux, CO_2 gas exchange, and thus photosynthetic activity [79]. Stomatal conductance changed by AM inoculation is closely connected to leaf water potential and relative water content in host plants. Under water restrictions, the first response of plants is stomatal closure to limit water loss through transpiration. Additionally, reduction of CO_2 uptake and carbon assimilation whereas favoring photorespiration may occur in plants [80]. Upregulation of *LeEPFL9* involved in the regulation of stomatal development together with greater stomatal density was found in tomato plants colonized by *R. irregularis* [10]. Inoculation of *Septoglomus deserticola* or *S. constrictum* sustained stomatal opening in host plants under drought conditions, substantially contributing to the carbon assimilation [11]. Improvement of stomatal conductance (g_s) in mycorrhizal castor bean [78], black locust [72], and strawberry [81] plants exposed to water stress has been detected.

One of the widely known benefits of mycorrhizal inoculation is the improvement of host water status under drought stress. Leaf water potential (LWP) and relative water content (RWC) of plants were substantially higher in the presence of mycorrhiza [11, 81]. Several studies illustrated a higher water use efficiency or intrinsic water use efficiency in AM plants during water stress [10, 81, 82]. It is believed that photosynthetic activity correlates with chlorophyll content and stomatal conductance, which have been enhanced by AMF. Drought stress changes photosynthetic pigments and damages chloroplasts. Nonetheless, AM inoculation alleviates the damage of these parameters caused by the stress [77]. *Rhizophagus irregularis*-colonized castor bean plants subjected to water restriction increased contents of chlorophyll a (by 26%), b (30%), carotenoid (by 28.5%), and total chlorophyll (25.5%) in comparison to counterparts of non-AM plants [78]. These increases in AM plants may be attributed to the improved nutrient uptake, particularly N and Mg that are structural components of chlorophyll.

Mycorrhizal colonization has been found to alleviate the adverse impacts of drought stress on photochemical efficiency and photosystem II (PSII) reaction center [77, 83]. Under water deficit, application of AMF promoted a higher maximum quantum efficiency of PSII (F_v/F_m) [11], greater photosynthetic efficiency [84], transpiration rate, and net photosynthesis rate (P_N) [10, 81]. Although mycorrhizal plants usually have higher photosynthetic capacities, environmental factors such as high atmospheric drought or low radiation can decide the beneficial effects of mycorrhiza on photosynthesis [85].

4.1.4 Phytohormonal changes

Phytohormones not only modulate a plethora of events during plant development but also are essential signaling molecules for interaction between plants and AMF [86]. Changes in plant hormone homeostasis also affect plant tolerance against abiotic stresses [87, 88]. During mycorrhization, changes in levels of several plant hormones have been reported [86], hence may contribute to the improved host plant tolerance to subsequent stresses.

Abscisic acid (ABA) is the most fundamental stress hormonal signal, modulating transpiration rate, root hydraulic conductivity, and aquaporin expression [89]. The concentration of ABA is heightened in plant tissues under drought stress to induce stomatal closure for reduction of water loss and activate different stress-responsive genes, increasing plant tolerance to drought [90]. A lower ABA concentration was found in roots and leaves of mycorrhizal plants versus nonmycorrhizal plants under drought stress [9, 10, 91]. Downregulation of *SlNCED* gene, a critical ABA biosynthetic gene, in *Septoglomus constrictum*-infected roots under water stress concurred with the greater g_s and higher water status of tomato plants, indicating a higher stress tolerance in colonized plants compared to uninoculated plants [11].

Nonetheless, an increase in ABA concentration in trifoliate orange plants colonized by *F. mosseae* was also observed under drought stress [73]. The reason for this remains poorly understood, which requires further research.

The role of jasmonate (JA) in water uptake and transport, exerting influence on stomatal conductance, root hydraulic conductance, and regulating the expression and abundance of aquaporins in tomato plants has been revealed [91]. Tomato plants defective in JA synthesis altered the AM impacts on the host plant, interfering phytohormones and expression of AM-induced aquaporin genes. The content of JA and its precursors was higher in leaves of *Digitaria eriantha* plants infected by *Rhizophagus irregularis* under water deficit, relative to noninfected plants, which could enhance plant tolerance to the stress [92]. Likewise, mycorrhizal inoculation substantially increased methyl jasmonate (MeJA) in trifoliate orange plants exposed to drought stress [93]. Under water-stress conditions, significantly higher expression levels of JA-biosynthetic gene *SlLOXD* in roots and leaves of colonized tomato plants were detected, supporting plant response to drought stress by triggering a LOXD-mediated pathway [10, 11].

Strigolactones (SLs), as phytohormones, not only modulate the coordinated development of plants exposed to nutrient shortages but are also host detection signals for AM establishment in the host plant [94]. Upregulation of the SL-biosynthesis gene *SlCCD7* together with a greater content of SLs was found in *Rhizophagus irregularis*-inoculated tomato roots subjected to water-stress conditions, correlated with the increase in AM colonization rate [9]. The stimulated production of SLs promoting symbiosis establishment as a strategy of plants to cope with drought stress has been proposed.

Auxin is a key regulator in root-hair initiation, growth, and developmental processes [95, 96]. In a recent study, an increased content of indole-3-acetic acid (IAA) which is the dominant naturally occurring auxin was found in mycorrhizal tomato plants exposed to drought [91]. Similarly, stimulation of biosynthesis and transport of IAA in roots of trifoliate orange under water restrictions were demonstrated [97]. Under drought conditions, AM colonization overexpressed *PtYUC3* and *PtYUC8* involved in IAA biosynthesis, and downregulated auxin efflux carriers (*PtPIN1* and *PtPIN3*), while up-regulated auxin-species influx carriers (*PtABCB19* and *PtLAX2*) in roots, leading to significantly higher IAA accumulation in mycorrhizal roots versus non-AM roots [97]. Together with higher IAA, colonized trifoliate orange plants showed a significant increase in MeJA, nitric oxide, and calmodulin in roots, supporting greater root adaptation of morphology as a crucial strategy for drought adaptation [93].

Although important roles of phytohormones are irrefragable in plant responses to water stress, little attention has been paid to them in mycorrhizal plants. Previous studies have just revealed changes in concentrations and expression of genes encoding biosynthesis of few hormones in colonized plants during drought stress; thereby, further research is required to understand it.

4.1.5 Osmotic adjustment

In response to drought stress, plants accumulate compatible solute compounds or osmolytes functioning for osmotic adjustment to maintain a favorable gradient for water uptake [98]. Osmotic adjustment is essential for water influx, turgor maintenance, sustaining physiological activity in plants such as stomatal opening, photosynthesis, cellular expansion, and growth during the stress [98]. Compatible solutes include a variety of sugars, proline, glycine betaine, polyamines, and organic acids such as oxalate and malate [99]. Interestingly, discrepant observations in osmolyte accumulation have been reported in a wide range of mycorrhizal plants [10, 83, 100, 101].

Proline, an amino acid, plays a crucial role in osmoregulation and acts as an efficient scavenger of reactive oxygen species (ROS) [102] (discussed in Section 4.1.7). Enhanced drought tolerance with a higher proline concentration in mycorrhizal plants has been shown in many studies [10, 78, 100]; nevertheless, opposite results have also been reported [81, 83]. Inoculation of either *F. mosseae* or *Paraglomus occultum* in trifoliate orange plants substantially reduced leaf proline content but improved the host plant growth under water deficit [103]. These results suggest that AMF strongly altered leaf proline metabolism through regulating proline-metabolized enzymes, which is important for osmotic adjustment of the host plants.

Sugars are osmoprotectants, which contribute up to 50% of osmotic potential in plants [104, 105]. In general, under water stress, the higher accumulation of total soluble sugars offers a defense mechanism in mycorrhizal plants such as watermelon [100] and flax [106]. Concentrations of sucrose, glucose, and fructose were significantly heightened in leaves of mycorrhizal trifoliate orange seedlings exposed to drought, which could function as osmolytes to stabilize and protect structures and macromolecules in plants from the stress, therefore improving host plant tolerance [103]. AMF-mediated increases in leaf sugar metabolism by modulating sugar-metabolized enzymes notably contribute to the osmotic adjustment of colonized plants. However, contrast observations have been shown in olive trees [101] and maize [107] colonized by AMF, which may be due to the fact that host plants suffer less stress. Noticeably, under severe drought inoculation with *Rhizophagus clarus* significantly reduced soluble sugars in leaves of strawberry plants, but this parameter was remarkably enhanced in roots in response to mild and severe water stress [81]. These changes together with an improved water status and plant biomass suggest different strategies for the enhanced water status triggered by AMF in roots and leaves of strawberry.

In summary, increased accumulation of compatible solutes in AM-inoculated plants in exposure to water deficit is supposed to protect plants from the stress and curtail the plant osmotic potential, whereas the lower osmolyte accumulation in host plants is thought to be due to colonized plants successfully gaining drought mitigation.

4.1.6 Enhanced plant tolerance to oxidative stress

One of the consequences of water stress is the overproduction of reactive oxygen species (ROS) such as hydroxyl radicals ($^{\cdot}OH$), superoxide radicals ($O_2^{\cdot-}$), singlet oxygen ($^1O_2$), and hydrogen peroxide (H_2O_2) mainly in chloroplasts and mitochondria. The excessive ROS results in unbalanced cellular homeostasis and then oxidative stress, damaging membrane lipids, proteins, and nucleic acids and even causing the death of cells [108]. To cope with oxidative stress, plants have evolved ROS scavengers in both nonenzymatic and enzymatic defense systems. Nonenzymatic antioxidants comprise phenolic compounds, glutathione, ascorbic acid, alkaloids, carotenoids, and tocopherol [109], which not only play a direct role in ROS removal but also serve as a substrate for the antioxidant enzymes in scavenging ROS [110]. Under water deficit, AMF ameliorate oxidative damage through augmented production of phenolic compounds and secondary metabolites detoxifying ROS in various plants [111–113]. AM inoculation also significantly increased the concentrations of anthocyanins and carotenoids [106] and ascorbic acid [82, 106] in plants in exposure to water constraints.

Another important ROS scavenger system is enzymatic antioxidants which could be enhanced in mycorrhizal plants including superoxide dismutase (SOD), catalase (CAT), peroxidase (POD), guaiacol peroxidase, ascorbate peroxidase (APX), glutathione reductase (GR), dehydroascorbate reductase (DHAR), and monodehydroascorbate reductase (MDHAR) [110]. The AM symbiosis has been reported to improve plant protection against oxidative stress by decreasing the level of lipid peroxidation (MDA) and H_2O_2 accumulation by strengthening significantly antioxidative enzymes SOD, POD, and CAT in roots and leaves under mild and severe drought [11, 81]. SOD and CAT are the most important ROS scavenging enzymes among the enzymatic antioxidants. These enzymes together with the cooperative enzymes (GR, MDHAR, DHAR, and APX) in the ascorbate-glutathione (ASA-GSH) cycle play pivotal roles in controlling overproduced ROS to maintain cellular homeostasis [114–116]. Remarkable increases in SOD, CAT, GR, APX, and MDHAR at transcription and enzymatic level correlated with lower $O_2^{\cdot-}$, H_2O_2, and MDA have been revealed in drought-stressed mycorrhizal plants versus counterparts of non-AM plants, improving host protection against oxidative damage [101].

Higher nonenzymatic and enzymatic antioxidants in colonized plants help for the rapid and efficient elimination of excess ROS. Nevertheless, discrepant results, no change or decrease in ROS scavengers, have also been demonstrated [70, 117]. Results are not entirely consistent with all reports because of different ages of host plants [118] and/or the specific combination of mycorrhizal strains and plant species, even cultivars [11] (as discussed in Section 3) or successful drought mitigation in colonized plants.

4.2 Indirect benefits of AM symbiosis for host plants under water deficit

The hyphal network of AMF is believed to improve soil water retention properties in the mycorrhizosphere through its physical, biological, and chemical influences. It has been reported that AMF produce polysaccharides, glomalin, mucilages, and hydrophobins that act to bind soil particles, leading to soil aggregation with enhanced water-holding capacity in soil [119]. Glomalin, a stable glycoprotein, highly persists in the soil, defined as glomalin-related soil protein (GRSP) [120]. The higher amounts of GRSP in the soil, the more enhanced capacity of water retention was found since soil aggregation increased protection of C-rich debris from the decomposition of soil microbes [120, 121]. Indeed, fungal hyphae coated by GRSP sharp a hydrophobic layer into the aggregate surface, hence decreasing water loss within soil aggregates [122]. When the fungal hyphae form branching structures with glomalin, they physically stick micro-aggregates with macro-aggregates [119]. The physical interaction of external hyphae on soil particles forms stable aggregates [123] in general and under water deficit [124]. Moreover, mycorrhizosphere also influences soil aggregation through alterations in the soil microbial food web, habitats for soil microbes, and biological activities in the host rhizosphere, which could result in an enhancement in microaggregate soil structure [125]. Thus, soils possess well-structured property in the presence of AMF, maintaining relatively higher available water than poorly structured soils without mycorrhizal presence under water stress [126]. Notably, in artificial substrates, an enhancement in water retention and water transport within substrates inoculated with AMF was observed under severe drought, suggesting that host plants perceive less stress at the root surface as reducing substrate moisture [127]. Hence, AMF postponed the physiological stress response in host plants.

5. Altered plant growth, yield, and quality

It is often found that AM symbiosis can improve plant growth in numerous plants, such as lettuce [9], tomato [9, 11], strawberry [81], maize [128], black locust [72], digitgrass (*Digitaria eriantha*), a source of forage [92], and damask rose [129].

The substantial improvement in the growth of mycorrhizal plants could be a result of a combination of AMF-induced mechanisms of plant tolerance under drought conditions, notably enhanced water and nutrient uptake in host plants [60, 117], and increased photosynthetic activity (as discussed in Section 4.1.3) since plant size closely links with measured physiological parameters [11]. It is important because nutrient supply may improve plant drought tolerance for better plant establishment. The increased plant biomass and nutrient uptake in AM plants could be more pronounced during seedling growth stages and in a longer stress duration. For instance, significant increases in shoot dry weight (by 128–242%) and root dry weight (185–328%) in French lavender (*Lavandula dentata*) plants treated by either single autochthonous AMF (*Septoglomus contrictum, Diversispora aunantia, Archaespora trappei, Glomus versiforme,* and *Paraglomus occultum*) or their mixture were recorded, compared to uninoculated plants after 6 months of growth under drought conditions [60].

Besides positive mycorrhizal effects on plant growth, discrepant observations have also been reported. Four tomato recombinant inbred lines (RIL 20, 40, 66 and 100) and one commercial cultivar inoculated with *Rhizophagus irregularis* showed variable results under water stress [37]. AM application remarkably increased shoot dry weight of RIL 40 and RIL 60 lines under drought conditions while no changes were recognized in plants colonized by other AMF. Similar results were found in soybean using single isolates of *Septoglomus constrictum*, *Glomus* sp., *Glomus aggregatum*, or their mixture [130]. Taken together, the benefits of AMF application under water deficit are dependent on the specific combination of plant genotypes and AM isolates.

Another significant benefit of mycorrhizal inoculation is to increase crop yield in exposure to water constraints compared to nonmycorrhizal plants. An array of observations shows a significantly higher yield, importantly marketable yield in mycorrhizal plants subjected to water scarcity in maize [128], tomato [82], flax [131], cowpea [132], and damask rose [129]. Furthermore, AM symbiosis has shown to accelerate flowering and fruit development [133]. Interestingly, re-inoculation of AMF after transplanting seedlings in the field appears to be necessary to strengthen mycorrhizal benefits. This could be seen in the field investigation which with twice application of AMF considerably heightened the marketable fruit yield (by 51–71%) in plants subjected to 50% water supply regime in comparison with those with mycorrhizal inoculation once at sowing and uninoculated ones [82]. The beneficial effect of AMF application on relative water content and nutritional status in plants as well as enhanced shoot accumulation of photoassimilates through higher photosynthetic activity, and improved stress tolerance in the presence of drought could result in higher productivity in colonized plants. Also, fruits are often the main sink for P; therefore, enhanced P nutrient in host plants promotes higher fruit yield.

As a result of physiological changes during mycorrhization, both transcriptional and metabolic changes occur in host plants influencing crop quality as well. AM symbiosis not only modulates gene expression in tomato fruit, through a systemic impact, but also changes the phenology of flowering and fruit ripening as well as in the amino acid profile [133]. Under water shortage, AMF treatment has been explored to improve quality attributes including antioxidant compounds, carotenoids and anthocyanins [82, 134], essential oils [135], and alteration in seed quality of flax [131], hence highlighting the potential of using AMF in crop production, producing industrial and oil plants.

6. Challenges of AMF application

Microbial symbionts of plants such as AMF represent a huge, but an unrealized resource for improving yields, especially in the tropics [136]; however, lower benefits to plants than the potential of these microorganisms are often found. To predict real benefits as well as all potentials of the fungal inoculation, implementation of field trials before AMF application on a large scale is indispensable in order to choose suitable inoculum or appropriately tune the best AM combination for target crop production systems. Moreover, various environmental factors influence the success of AMF application into the field.

Another critical issue is whether generic or tuned AM products should be utilized in sustainable crop production. One of the challenges of AMF inoculation under open field conditions is the native populations of AMF in soils, which are able to remarkably compete colonization niche with the introduced symbionts. Despite the fact that commercial AMF inoculants are usually advertised as compatible for a variety of host plants and field-cultivation conditions, the AM-induced benefits for crops are not always as expected [137]. The bridge between research and AM suppliers should be strengthened to recommend appropriate AM inocula for most benefits. Due to the specificity of AMF-host plant interaction as described in several places in the chapter, an attempt to exploit advantageous combinations is necessary. Fine-tuning commercial mycorrhizal products is vital to obtain optimum beneficial effects from mycorrhizal inoculation.

Even in some circumstances, the symbiotic effectiveness and adaptability of the indigenous fungi are more dominant than non-native ones [138, 139]; therefore, introduced AMF isolates could be less profitable than native ones [140]. Besides, there is an existence of functional diversity among different AMF species [36]. Remarkable differences in performance even among different geographical isolates belonging to the same mycorrhizal species have been described [141]. In such cases, isolation of indigenous mycorrhizal strains for inoculum production, then large-scale reintroduction of these native fungi in the field could be a feasible solution for a useful AMF application [142]. It is worth mentioning that selection of specific AM taxa for particular crops is the best approach to improve crop growth, and there is no 'one-size-fits-all' AMF [143]. In controlled environments, application of a single AMF is more effective than using a mixture of different AM taxa [143].

During the last decades, several molecular techniques have been used to characterize entire communities of mycorrhiza in soil [144, 145] and AMF inocula [146–148]. These techniques enable to monitor the introduced fungal symbiont both inside and outside the host during plant growth [149, 150]. Tracing the introduced AMF temporally and spatially could be implemented by high-throughput next-generation sequencing, which possibly verifies whether the introduced fungi favor substantial levels of colonization and explores how the inoculated AMF coexist and interact with the local community of AMF [136]. Advances in molecular techniques can further assist the adjustment or tuning of commercial inoculants to specific AMF combinations with host plants under crop production systems.

Another major limitation of mycorrhizal inoculation in horticulture and agriculture is farmer's awareness and acceptance and the relatively high cost of it. Furthermore, conventional breeding programs have overlooked plant characteristics facilitating mycorrhizal association, and plant breeders have selected varieties in favor of acquiring nutrients in high-input crop production systems without respect to the AMF role in soil nutrient management [151], resulting in the primary challenge to AMF application. Hence, modern breeding programs should consider AMF as an essential component of breeding traits in new cultivars, particularly those cultivated under environmental adversities such as drought stress in which AMF application has been proved to stimulate higher crop tolerance.

7. Conclusion and future perspectives

AM inoculation can offer multiple advantages to host plants in exposure to water scarcity, which could enable inoculated plants to avoid drought stress or tolerate water deficit better than nonmycorrhizal plants. Indeed, various direct and indirect AM-induced mechanisms in mycorrhizal plants could contribute to drought mitigation or tolerance. More importantly, improved crop yield and quality attribute in colonized plants under drought stress highlight the importance of AMF application in crop production as one of the promising practices under water constraints. However, variable plant responses to AMF and the discussed major challenges hinder possible fruitful outcomes of AM inoculation. Identification of the most appropriate combination of fungal inoculants and a given variety, cultivar, or accession grown under water scarcity, and understanding environmental factors deciding the positive results of the inoculation are crucial determinants for successful AMF application. Compatible combination of AMF with other beneficial microbes such as plant growth-promoting bacteria and/or *Trichoderma* offering synergistic effects on plant tolerance to stressful environments including drought stress is also a bright perspective [38, 106]. Besides that, further research is necessary to shed light on the specific functions of genes mediated by mycorrhiza, which could explore the exact AM-triggered mechanisms of plant adaptation under water deficit. Studies on quantitative trait loci (QTL) involved in mycorrhizal plant responses to drought stress are needed for breeding programs to create new cultivars with a combination of drought-tolerant traits and AM benefits.

Acknowledgements

This work was supported by 1783-3/2018/FEKUTSTRAT Program awarded by the Ministry of Human Capacities and by the National Research Development and Innovation Office (2017-1.3.1-VKE-2017-00022).

Conflict of interest

We declare that we do not have any conflict of interest.

Author details

Katalin Posta* and Nguyen Hong Duc
Institute of Genetics, Microbiology and Biotechnology, Szent Istvan University, Hungary

*Address all correspondence to: posta.katalin@mkk.szie.hu

IntechOpen

© 2019 The Author(s). Licensee IntechOpen. This chapter is distributed under the terms of the Creative Commons Attribution License (http://creativecommons.org/licenses/by/3.0), which permits unrestricted use, distribution, and reproduction in any medium, provided the original work is properly cited.

References

[1] Sheffield J, Wood EF, Roderick ML. Little change in global drought over the past 60 years. Nature. 2012;**491**:435-438. DOI: 10.1038/nature11575

[2] Lambers H, Chapin FS, Pons TL. Plant Physiological Ecology. 2nd ed. New York: Springer; 2008

[3] Osakabe Y, Osakabe K, Shinozaki K, Tran L-SP. Response of plants to water stress. Frontiers in Plant Science. 2014;**5**:86. DOI: 10.3389/fpls.2014.00086

[4] Daryanto S, Wang L, Jacinthe PA. Global synthesis of drought effects on maize and wheat production. PLoS One. 2016;**11**:e0156362. DOI: 10.1371/journal.pone.0156362

[5] Ngumbi E, Kloepper J. Bacterial-mediated drought tolerance: Current and future prospects. Applied Soil Ecology. 2014;**105**:109-125. DOI: 10.1016/j.apsoil.2016.04.009

[6] Bonfante P, Genre A. Mechanisms underlying beneficial plant-fungus interactions in mycorrhizal symbiosis. Nature Communications. 2010;**1**:48. DOI: 10.1038/ncomms1046

[7] Balestrini R, Lumini E. Focus on mycorrhizal symbioses. Applied Soil Ecology. 2018;**123**:299-304. DOI: 10.1016/j.apsoil.2017.09.001

[8] Birhane E, Sterck FJ, Fetene M, Bongers F, Kuyper TW. Arbuscular mycorrhizal fungi enhance photosynthesis, water use efficiency, and growth of frankincense seedlings under pulsed water availability conditions. Oecologia. 2012;**169**(4):895-904. DOI: 10.1007/s00442-012-2258-3

[9] Ruiz-Lozano JM, Aroca R, Zamarreno AM, Molina S, Andreo-Jiménez B, Porcel R, et al. Arbuscular mycorrhizal symbiosis induces strigolactone biosynthesis under drought and improves drought tolerance in lettuce and tomato. Plant, Cell & Environment. 2015;**39**:441-452. DOI: 10.1111/pce.12631

[10] Chitarra W, Pagliarani C, Maserti B, Lumini E, Siciliano I, Cascone P, et al. Insights on the impact of arbuscular mycorrhizal symbiosis on tomato tolerance to water stress. Plant Physiology. 2016;**171**:1009-1023. DOI: 10.1104/pp.16.00307

[11] Duc NH, Csintalan Z, Posta K. Arbuscular mycorrhizal fungi mitigate negative effects of combined drought and heat stress on tomato plants. Plant Physiology and Biochemistry. 2018;**132**:297-307. DOI: 10.1016/j.plaphy.2018.09.011

[12] Smith SE, Read DJ. Mycorrhizal Symbiosis. 3rd ed. San Diego: Academic Press; 2008

[13] Allen MF, Swenson W, Querejeta JI, Egerton-Warburton LM, Treseder KK. Ecology of mycorrhizae: A conceptual framework for complex interactions among plants and fungi. Annual Review of Phytopathology. 2003;**41**:271-303. DOI: 10.1146/annurev.phyto.41.052002.095518

[14] Bravo A, Brands M, Wewer V, Dörmann P, Harrison MJ. Arbuscular mycorrhiza-specific enzymes FatM and RAM2 fine-tune lipid biosynthesis to promote development of arbuscular mycorrhiza. New Phytologist. 2017;**214**(4):1631-1645. DOI: 10.1111/nph.14533

[15] Olson PA, Thingstrub I, Jakobsen I, Baath E. Estimation of the biomass of arbuscular mycorrhizal fungi in a linseed field. Soil Biology and Biochemistry. 1999;**31**(13):1879-1887. DOI: 10.1016/S0038-0717(99)00119-4

[16] Lu XH, Koide RT. The effect of mycorrhizal infection on components of plant-growth and reproduction. New Phytologist. 1994;**128**(2):211-218. DOI: 10.1111/j.1469-8137.1994.tb04004.x

[17] Duc NH, Posta K. Mycorrhiza-induced alleviation of plant disease caused by *Clavibacter michiganensis* subsp. *michiganensis* and role of ethylene in mycorrhiza-induced resistance in tomato. Acta Biologica Hungarica. 2018;**69**(2):170-181. DOI: 10.1556/018.69.2018.2.6

[18] Redecker D, Schüßler A, Stockinger H, Stürmer SL, Morton JB, Walker C. An evidence-based consensus for the classification of arbuscular mycorrhizal fungi (Glomeromycota). Mycorrhiza. 2013;**23**(7):515-531. DOI: 10.1007/s00572-013-0486-y

[19] Krüger M, Krüger C, Walker C, Stockinger H, Schüßler A. Phylogenetic reference data for systematics and phylotaxonomy of arbuscular mycorrhizal fungi from phylum to species level. New Phytologist. 2012;**193**(4):970-984. DOI: 10.1111/j.1469-8137.2011.03962.x

[20] Chen ECH, Morin E, Beaudet D, Noel J, Yildirir G, Ndikumana S, et al. High intraspecific genome diversity in the model arbuscular mycorrhizal symbiont *Rhizophagus irregularis*. New Phytologist. 2018;**220**(4):1161-1171. DOI: 10.1111/nph.14989

[21] Bever JD, Wang M. Arbuscular mycorrhizal fungi: Hyphal fusion and multigenomic structure. Nature. 2005;**433**:E3-E4. DOI: 10.1038/nature03294

[22] Chabaud M, Genre A, Sieberer BJ, Faccio A, Fournier J, Novero M, et al. Arbuscular mycorrhizal hyphopodia and germinated spore exudates trigger Ca2+ spiking in the legume and nonlegume root epidermis. New Phytologist. 2011;**189**:347-355. DOI: 10.1111/j.1469-8137.2010.03464.x

[23] Goltapeh EM, Danesh YR, Prasad R, Varma A. Mycorrhizal fungi: What we know and what should we know? In: Varma A, editor. Mycorrhiza. 3rd ed. Berlin Heidelberg: Springer-Verlag; 2008. pp. 3-27

[24] Morton JB, Benny GL. Revised classification of arbuscular mycorrhizal fungi (Zygomycetes): A new order, Glomales, two new suborders, Glomineae and Gigasporineae, and two new families, Acaulosporaceae and Gigasporaceae, with an emendation of Glomaceae. Mycotaxon. 1990;**37**:471-491

[25] Gutjahr C, Parniske M. Cell and developmental biology of arbuscular mycorrhiza symbiosis. Annual Review of Cell and Developmental Biology. 2013;**29**:593-617. DOI: 10.1146/annurev-cellbio-101512-122413

[26] Abbott LK, Robinson AD. Formation of external hyphae in soil by four species of vesicular-arbuscular mycorrhizal fungi. New Phytologist. 1985;**99**:245-255. DOI: 10.1111/j.1469-8137.1985.tb03653.x

[27] Giovannetti M, Fortuna P, Citernesi AS, Morini S, Nuti MP. The occurrence of anastomosis formation and nuclear exchange in intact arbuscular mycorrhizal networks. New Phytologist. 2001;**151**:717-724. DOI: 10.1046/j.0028-646x.2001.00216.x

[28] Hart MM, Reader RJ. Taxonomic basis for variation in the colonization strategy of arbuscular mycorrhizal fungi. New Phytologist. 2002;**153**:335-344. DOI: 10.1046/j.0028-646X.2001.00312.x

[29] Miransari M. Signaling molecules in the arbuscular mycorrhizal fungi. In: Gupta VK, Ayyachamy M, editors. Biotechnology of Fungal Genes. Boca Raton: Taylor & Francis Group, LLC; 2012. pp. 245-263

[30] Cosme M, Fernández I, Van der Heijden MGA, Pieterse CMJ. Nonmycorrhizal plants: The exceptions that prove the rule. Trends in Plant Science. 2018;**23**(7). DOI: 10.1016/j.tplants.2018.04.004

[31] Yoneyama K, Xie X, Sekimoto H, Takeuchi Y, Ogasawara S, Akiyama K, et al. Strigolactones, host recognition signals for root parasitic plants and arbuscular mycorrhizal fungi, from Fabaceae plants. New Phytologist. 2008;**179**:484-494. DOI: 10.1111/j.1469-8137.2008.02462.x

[32] Jin L, Wang Q, Wang Q, Wang X, Gange AC. Mycorrhizal-induced growth depression in plants. Symbiosis. 2017;**72**:81-88. DOI: 10.1007/s13199-016-0444-5

[33] Grunwald U, Guo W, Fischer K, Isayenkov S, Ludwig-Müller J, Hause B, et al. Overlapping expression patterns and differential transcript levels of phosphate transporter genes in arbuscular mycorrhizal, pi-fertilised and phytohormone-treated *Medicago truncatula* roots. Planta. 2009;**229**:1023-1034. DOI: 10.1007/s00425-008-0877-z

[34] Schmidt B, Gaşpar S, Camen D, Ciobanu I, Sumălan R. Arbuscular mycorrhizal fungi in terms of symbiosis-parasitism continuum. Communications in Agricultural and Applied Biological Sciences. 2011;**76**(4):653-659

[35] Fitter AH. What is the link between carbon and phosphorus fluxes in arbuscular mycorrhizas? A null hypothesis for symbiotic function. New Phytologist. 2006;**172**:3-6. DOI: 10.1111/j.1469-8137.2006.01861.x

[36] Rivero J, Alvarez D, Flors V, Azcón-Aguilar C, Pozo MJ. Root metabolic plasticity underlies functional diversity in mycorrhiza-enhanced stress tolerance in tomato. New Phytologist. 2018;**220**:1322-1336. DOI: 10.1111/nph.15295

[37] Calvo-Polanco M, Sánchez-Romer B, Aroca R, Asins MJ, Declerck S, Dodd IC, et al. Exploring the use of recombinant inbred lines in combination with beneficial microbial inoculants (AM fungus and PGPR) to improve drought stress tolerance in tomato. Environmental and Experimental Botany. 2016;**131**:47-57. DOI: 10.1016/j.envexpbot.2016.06.015

[38] Duc NH, Mayer Z, Pék Z, Helyes L, Posta K. Combined inoculation of arbuscular mycorrhizal fungi, *Pseudomonas fluorescens* and *Trichoderma* spp. for enhancing defense enzymes and yield of three pepper cultivars. Applied Ecology and Environmental Research. 2017;**15**(3):1815-1829. DOI: 10.15666/aeer/1503_18151829

[39] Hetrick BAD, Wilson GWT, Cox TS. Mycorrhizal dependence of modern wheat cultivars and ancestors: A synthesis. Canadian Journal of Botany. 1993;**71**:512-518. DOI: 10.1139/b93-056

[40] An G-H, Kobayashi S, Enoki H, Sonobe K, Muraki M, Karasawa T, et al. How does arbuscular mycorrhizal colonization vary with host plant genotype? An example based on maize (*Zea mays*) germplasms. Plant and Soil. 2010;**327**:441-453. DOI: 10.1007/s11104-009-0073-3

[41] Estaún V, Calvet C, Camprubí A. Effect of differences among crop species and cultivars on the arbuscular mycorrhizal symbiosis. In: Koltai H, Kapulnik Y, editors. Arbuscular Mycorrhizas: Physiology and Function. 2nd ed. New York: Springer Science+Business Media B.V; 2010. pp. 279-295. DOI: 10.1007/978-90-481-9489-6_13

[42] Bazghaleh N, Hamel C, Gan Y, Tar'an B, Knight JD. Genotypic variation in the response of chickpea to arbuscular

mycorrhizal fungi and non-mycorrhizal fungal endophytes. Canadian Journal of Microbiology. 2018;**64**(4):265-275. DOI: 10.1139/cjm-2017-0521

[43] Watts-Williams SJ, Cavagnaro TR, Tyerman SD. Variable effects of arbuscular mycorrhizal fungal inoculation on physiological and molecular measures of root and stomatal conductance of diverse *Medicago truncatula* accessions. Plant, Cell & Environment. 2019;**42**:285-294. DOI: 10.1111/pce.13369

[44] Martin-Robles N, Lehmann A, Seco E, Aroca R, Rillig MC, Milla R. Impacts of domestication on the arbuscular mycorrhizal symbiosis of 27 crop species. New Phytologist. 2018;**218**:322-334. DOI: 10.1111/nph.14962

[45] Marulanda A, Azcón R, Ruiz-Lozano JM. Contribution of six arbuscular mycorrhizal fungal isolates to water uptake by *Lactuca sativa* plants under drought stress. Physiologia Plantarum. 2003;**119**:526-533. DOI: 10.1046/j.1399-3054.2003.00196.x

[46] Dodd JC, Boddington CL, Rodriguez A, Gonzalez-Chavez C, Mansur I. Mycelium of arbuscular mycorrhizal fungi (AMF) from different genera: Form, function and detection. Plant and Soil. 2000;**226**(2):131-151. DOI: 10.1023/A:1026574828169

[47] Allen MF. Influence of vesicular-arbuscular mycorrhizae on water movement through *Bouteloua gracilis* lag ex Steud. New Phytologist. 1982;**91**:191-196. DOI: 10.1111/j.1469-8137.1982.tb03305.x

[48] Faber BA, Zasoski RJ, Munns DN. A method for measuring hyphal nutrient and water uptake in mycorrhizal plants. Canadian Journal of Botany. 1991;**69**:87-94. DOI: 10.1139/b91-012

[49] Ruth B, Khalvati M, Schmidhalter U. Quantification of mycorrhizal water uptake via high-resolution on-line water content sensors. Plant and Soil. 2011;**342**:459-468. DOI: 10.1007/s11104-010-0709-3

[50] Smith SE, Smith FA. Roles of arbuscular mycorrhizas in plant nutrition and growth: New paradigms from cellular to ecosystem scales. Annual Review of Plant Biology. 2011;**62**:227-250. DOI: 10.1146/annurev-arplant-042110-103846

[51] Volpe V, Giovannetti M, Sun X-G, Fiorilli V, Bonfante P. The phosphate transporters LjPT4 and MtPT4 mediate early root responses to phosphate status in nonmycorrhizal roots. Plant, Cell & Environment. 2015;**39**(3):660-671. DOI: 10.1111/pce.12659

[52] Hu W, Zhang H, Zhang X, Chen H, Tang M. Characterization of six PHT1 members in *Lycium barbarum* and their response to arbuscular mycorrhiza and water stress. Tree Physiology. 2017;**37**:351-366. DOI: 10.1093/treephys/tpw125

[53] Volpe V, Chitarra W, Cascone P, Volpe MG, Bartolini P, Moneti G, et al. The association with two different arbuscular mycorrhizal fungi differently affects water stress tolerance in tomato. Frontiers in Plant Science. 2018;**9**:1480. DOI: 10.3389/fpls.2018.01480

[54] Kikuchi Y, Hijikata N, Ohtomo R, Handa Y, Kawaguchi M, Saito K, et al. Aquaporin-mediated long-distance polyphosphate translocation directed towards the host in arbuscular mycorrhizal symbiosis: Application of virus-induced gene silencing. New Phytologist. 2016;**211**:1202-1208. DOI: 10.1111/nph.14016

[55] Boomsma CR, Vyn TJ. Maize drought tolerance: Potential improvements through arbuscular mycorrhizal symbiosis? Field Crops Research. 2008;**108**:14-31. DOI: 10.1016/j.fcr.2008.03.002

[56] Cardoso IM, Kuyper TW. Mycorrhizas and tropical soil fertility. Agriculture, Ecosystems and Environment. 2006;**116**:72-84. DOI: 10.1016/j.agee.2006.03.011

[57] Tisserant E, Kohler A, Dozolme-Seddas P, Balestrini R, Benabdellah K, Colard A, et al. The transcriptome of the arbuscular mycorrhizal fungus *Glomus intraradices* (DAOM197198) reveals functional tradeoffs in an obligate symbiont. New Phytologist. 2012;**193**:755-769. DOI: 10.1111/j.1469-8137.2011.03948.x

[58] Tamayo E, Gómez-Gallego T, Azcón-Aguilar C, Ferrol N. Genome-wide analysis of copper, iron and zinc transporters in the arbuscular mycorrhizal fungus *Rhizophagus irregularis*. Plant Traffic and Transport. 2014;**5**:547. DOI: 10.3389/fpls.2014.00547

[59] Kobae Y, Tamura Y, Takai S, Banba M, Hata S. Localized expression of arbuscular mycorrhiza-inducible ammonium transporters in soybean. Plant & Cell Physiology. 2010;**51**:1411-1415. DOI: 10.1093/pcp/pcq099

[60] Koegel S, Ait Lahmidi N, Arnould C, Chatagnier O, Walder F, Ineichen K, et al. The family of ammonium transporters (AMT) in *Sorghum bicolor*: Two AMT members are induced locally, but not systemically in roots colonized by arbuscular mycorrhizal fungi. New Phytologist. 2013;**198**:853-865. DOI: 10.1111/nph.12199

[61] Armada E, Probanza A, Roldán A, Azcón R. Native plant growth promoting bacteria *Bacillus thuringiensis* and mixed or individual mycorrhizal species improved drought tolerance and oxidative metabolism in *Lavandula dentata* plants. Journal of Plant Physiology. 2016;**192**:1-12. DOI: 10.1016/j.jplph.2015.11.007

[62] Attarzadeh M, Balouchi H, Rajaie M, Dehnavi MM, Salehi A. Growth and nutrient content of *Echinacea purpurea* as affected by the combination of phosphorus with arbuscular mycorrhizal fungus and *Pseudomonas florescent* bacterium under different irrigation regimes. Journal of Environmental Management. 2019;**231**:182-188. DOI: 10.1016/j.jenvman.2018.10.040

[63] Gholamhoseini M, Ghalavand A, Dolatabadian A, Jamshidi E, Khodaei-Joghan A. Effects of arbuscular mycorrhizal inoculation on growth, yield, nutrient uptake and irrigation water productivity of sunflowers grown under drought stress. Agricultural Water Management. 2013;**117**:106-114. DOI: 10.1016/j.agwat.2012.11.007

[64] Maurel C, Boursiac Y, Luu D-T, Santoni V, Shahzad Z, Verdoucq L. Aquaporins in plants. Physiological Reviews. 2015;**95**:1321-1358. DOI: 10.1152/physrev.00008.2015

[65] Diehn TA, Pommerrenig B, Bernhardt N, Hartmann A, Bienert GP. Genome-wide identification of aquaporin encoding genes in *Brassica oleracea* and their phylogenetic sequence comparison to *Brassica* crops and *Arabidopsis*. Frontiers in Plant Science. 2015;**6**:166. DOI: 10.3389/fpls.2015.00166

[66] Aroca R, Bago A, Sutka M, Paz JA, Cano C, Amodeo G, et al. Expression analysis of the first arbuscular mycorrhizal fungi aquaporin described reveals concerted gene expression between salt-stressed and nonstressed mycelium. Molecular Plant-Microbe Interactions. 2009;**22**:1169-1178. DOI: 10.1094/MPMI-22-9-1169

[67] Li T, Hu YJ, Hao ZP, Li H, Wang YS, Chen BD. First cloning and characterization of two functional aquaporin genes from an arbuscular mycorrhizal fungus *Glomus intraradices*. New Phytologist. 2013;**197**:617-630. DOI: 10.1111/nph.12011

[68] Li T, Hu YJ, Hao ZP, Li H, Chen BD. Aquaporin genes *GintAQPF1* and *GintAQPF2* from *Glomus intraradices* contribute to plant drought tolerance. Plant Signaling & Behavior. 2013;**8**(5):e24030. DOI: 10.4161/psb.24030

[69] Bárzana G, Aroca R, Bienert GP, Chaumont F, Ruiz-Lozano JM. New insights into the regulation of aquaporins by the arbuscular mycorrhizal symbiosis in maize plants under drought stress and possible implications for plant performance. Molecular Plant-Microbe Interactions. 2014;**27**:349-363. DOI: 10.1094/MPMI-09-13-0268-R

[70] Armada E, Azcon R, Lopez-Castillo OM, Calvo-Polanco M, Ruiz-Lozano JM. Autochthonous arbuscular mycorrhizal fungi and *Bacillus thuringiensis* from a degraded Mediterranean area can be used to improve physiological traits and performance of a plant of agronomic interest under drought conditions. Plant Physiology and Biochemistry. 2015;**90**:64-74. DOI: 10.1016/j.plaphy.2015.03.004

[71] Quiroga G, Erice G, Aroca R, Chaumont F, Ruiz-Lozano JM. Enhanced drought stress tolerance by the arbuscular mycorrhizal symbiosis in a drought-sensitive maize cultivar is related to a broader and differential regulation of host plant aquaporins than in a drought-tolerant cultivar. Frontiers in Plant Science. 2017;**8**:1056. DOI: 10.3389/fpls.2017.01056

[72] He F, Zhang H, Tang M. Aquaporin gene expression and physiological responses of *Robinia pseudoacacia* L. to the mycorrhizal fungus *Rhizophagus irregularis* and drought stress. Mycorrhiza. 2016;**26**(4):311-323. DOI: 10.1007/s00572-015-0670-3

[73] He J-D, Dong T, Wu H-H, Ying-Ning Z, Qiang-Sheng W, Kuča K. Mycorrhizas induce diverse responses of root *TIP* aquaporin gene expression to drought stress in trifoliate orange. Scientia Horticulturae. 2019;**243**:64-69. DOI: 10.1016/j.scienta.2018.08.010

[74] Calvo-Polanco M, Sánchez-Castro I, Cantos M, García JL, Azcón R, Ruiz-Lozano JM, et al. Effects of different arbuscular mycorrhizal fungal backgrounds and soils on olive plants growth and water relation properties under well-watered and drought conditions. Plant, Cell & Environment. 2016;**39**:2498-2514. DOI: 10.1111/pce.12807

[75] Liu T, Li Z, Hui C, Tang M, Zhang H. Effect of *Rhizophagus irregularis* on osmotic adjustment, antioxidation and aquaporin PIP genes expression of *Populus* × *canadensis* 'Neva' under drought stress. Acta Physiologiae Plantarum. 2016;**38**(8):191-191. DOI: 10.1007/s11738-016-2207-6

[76] Recchia GH, Konzen ER, Cassieri F, Caldas DGG, Tsai SM. Arbuscular mycorrhizal symbiosis leads to differential regulation of drought-responsive genes in tissue-specific root cells of common bean. Frontiers in Microbiology. 2018;**9**:1339. DOI: 10.3389/fmicb.2018.01339

[77] Mathur S, Tomar RS, Jajoo A. Arbuscular mycorrhizal fungi (AMF) protects photosynthetic apparatus of wheat under drought stress. Photosynthesis Research. 2019;**139**:227-238. DOI: 10.1007/s11120-018-0538-4

[78] Zhang T, Hub Y, Zhang K, Tian C, Guo J. Arbuscular mycorrhizal fungi improve plant growth of *Ricinus communis* by altering photosynthetic properties and increasing pigments under drought and salt stress. Industrial Crops and Products. 2018;**117**:13-19. DOI: 10.1016/j.indcrop.2018.02.087

[79] Augé RM, Toler HD, Saxton AM. Arbuscular mycorrhizal symbiosis alters stomatal conductance of host plants

more under drought than under amply watered conditions: A meta-analysis. Mycorrhiza. 2015;**25**(1):13-24. DOI: 10.1007/s00572-014-0585-4

[80] Samarah NH, Alqudah AM, Amayreh JA, McAndrews GM. The effect of late-terminal drought stress on yield components of four barley cultivars. Journal of Agronomy and Crop Science. 2009;**195**:427-441. DOI: 10.1111/j.1439-037X.2009.00387.x

[81] Moradtalab N, Hajiboland R, Aliasgharzad N, Hartmann TE, Neumann G. Silicon and the association with an arbuscular-mycorrhizal fungus (*Rhizophagus clarus*) mitigate the adverse effects of drought stress on strawberry. Agronomy. 2019;**9**:41. DOI: 10.3390/agronomy9010041

[82] Bakr J, Daood HG, Pék Z, Helyes L, Posta K. Yield and quality of mycorrhized processing tomato under water scarcity. Applied Ecology and Environmental Research. 2017;**15**(1):401-413. DOI: 10.15666/aeer/1501_401413

[83] Bakr J, Pék Z, Helyes L, Posta K. Mycorrhizal inoculation alleviates water deficit impact on field-grown processing tomato. Polish Journal of Environmental Studies. 2018;**27**(5):1949-1958. DOI: 10.15244/pjoes/78624

[84] Ruiz-Sanchez M, Aroca R, Muñoz Y, Polón R, Ruiz-Lozano JM. The arbuscular mycorrhizal simbiosis enhances the photosynthetic efficiency and the antioxidative response of rice plants subjected to drought stress. Journal of Plant Physiology. 2010;**167**:862-869. DOI: 10.1016/j.jplph.2010.01.018

[85] Bitterlich M, Franken P, Graefe J. Atmospheric drought and low light impede mycorrhizal effects on leaf photosynthesis—A glasshouse study on tomato under naturally fluctuating environmental conditions. Mycorrhiza. 2019;**29**:13-28. DOI: 10.1007/s00572-018-0872-6

[86] Ludwig-Muller J. Hormonal responses in host plants triggered by arbuscular mycorrhizal fungi. In: Kapulnik Y, Koltai H, editors. Arbuscular Mycorrhizas: Physiology and Function. 2nd ed. Dordrecht: Springer; 2010. pp. 169-190

[87] Pineda A, Dicke M, Pieterse CMJ, Pozo MJ. Beneficial microbes in a changing environment: Are they always helping plants to deal with insects? Functional Ecology. 2013;**27**(3):574-586. DOI: 10.1111/1365-2435.12050

[88] Ruiz-Lozano JM, Porcel R, Azcón-Aguilar C, Aroca R. Regulation by arbuscular mycorrhiza of the integrated physiological response to salinity in plants: New challenges in physiological and molecular studies. Journal of Experimental Botany. 2012;**63**:4033-4044. DOI: 10.1093/jxb/ers126

[89] Aroca R, Porcel R, Ruiz-Lozano JM. How does arbuscular mycorrhizal symbiosis regulate root hydraulic properties and plasma membrane aquaporins in *Phaseolus vulgaris* under drought, cold or salinity stresses? New Phytologist. 2007;**173**:808-816. DOI: 10.1111/j.1469-8137.2006.01961.x

[90] Lim CW, Baek W, Jung J, Kim J-H, Lee SC. Function of ABA in stomatal defense against biotic and drought stresses. International Journal of Molecular Sciences. 2015;**16**:15251-15270. DOI: 10.3390/ijms160715251

[91] Sánchez-Romera B, Calvo-Polanco M, Ruiz-Lozano JM, Zamarreño AM, Arbona V, García-Mina JM, et al. Involvement of *def-1* mutation in the response of tomato plants to arbuscular mycorrhizal symbiosis under well watered and drought conditions. Plant & Cell Physiology. 2018;**59**:248-261. DOI: 10.1093/pcp/pcx178

[92] Pedranzani H, Rodríguez-Rivera M, Gutiérrez M, Porcel R, Hause B, Ruiz-Lozano JM. Arbuscular mycorrhizal symbiosis regulates physiology and performance of *Digitaria eriantha* plants subjected to abiotic stresses by modulating antioxidant and jasmonate levels. Mycorrhiza. 2016;**26**:141-152. DOI: 10.1007/s00572-015-0653-4

[93] Zou Y-N, Wang P, Liu C-Y, Ni Q-D, Zhang D-J, Wu Q-S. Mycorrhizal trifoliate orange has greater root adaptation of morphology and phytohormones in response to drought stress. Scientific Reports. 2017;**7**:41134. DOI: 10.1038/srep41134

[94] López-Ráez JA. How drought and salinity affect arbuscular mycorrhizal symbiosis and strigolactone biosynthesis? Planta. 2016;**243**:1375-1385. DOI: 10.1007/s00425-015-2435-9

[95] Zhang DJ, Xia RX, Cao X. Ethylene modulates root hair development in trifoliate orange through auxin-signaling pathway. Scientia Horticulturae. 2016;**213**:252-259. DOI: 10.1016/j.scienta.2016.11.007

[96] Leyser O. Auxin signaling. Plant Physiology. 2018;**176**:465-479. DOI: 10.1104/pp.17.0076

[97] Liu C-Y, Zhang F, Zhang D-J, Srivastava AK, Wu Q-S, Zou Y-N. Mycorrhiza stimulates root-hair growth and IAA synthesis and transport in trifoliate orange under drought stress. Scientific Reports. 2018;**8**:1978. DOI: 10.1038/s41598-018-20456-4

[98] Ruiz-Lozano JM. Arbuscular mycorrhizal symbiosis and alleviation of osmotic stress. New perspectives for molecular studies. Mycorrhiza. 2003;**13**(6):309-317. DOI: 10.1007/s00572-003-0237-6

[99] Singh M, Kumar J, Singh S, Singh VP, Prasad SM. Roles of osmoprotectants in improving salinity and drought tolerance in plants: A review. Reviews in Environmental Science and Biotechnology. 2015;**14**:407-426. DOI: 10.1007/s11157-015-9372-8

[100] Mo Y, Wang Y, Yang R, Zheng J, Liu C, Li H, et al. Regulation of plant growth, photosynthesis, antioxidation and osmosis by an arbuscular mycorrhizal fungus in watermelon seedlings under well-watered and drought conditions. Frontiers in Plant Science. 2016;**7**:644. DOI: 10.3389/fpls.2016.00644

[101] Ouledali S, Ennajeh M, Zrig A, Gianinazzi S, Khemira H. Estimating the contribution of arbuscular mycorrhizal fungi to drought tolerance of potted olive trees (*Olea europaea*). Acta Physiologiae Plantarum. 2018;**40**:81. DOI: 10.1007/s11738-018-2656-1

[102] Hayat S, Hayat Q, Alyemeni MN, Wani AS, Pichtel J, Ahmad A. Role of proline under changing environments: A review. Plant Signaling & Behavior. 2012;**7**(11):1456-1466. DOI: 10.4161/psb.21949

[103] Wu H-H, Zou Y-N, Rahman MM, Ni Q-D, Wu Q-S. Mycorrhizas alter sucrose and proline metabolism in trifoliate orange exposed to drought stress. Scientific Reports. 2017;**7**:42389. DOI: 10.1038/srep42389

[104] Abdel-Latef A, Chaoxing H. Does the inoculation with *Glomus mosseae* improve salt tolerance in pepper plants? Journal of Plant Growth Regulation. 2014;**33**(3):644-653. DOI: 10.1007/s00344-014-9414-4

[105] Sheng M, Tang M, Zhang F, Huang Y. Influence of arbuscular mycorrhiza on organic solutes in maize leaves under salt stress. Mycorrhiza. 2011;**21**:423-430. DOI: 10.1007/s00572-010-0353-z

[106] Rahimzadeh S, Pirzad A. Arbuscular mycorrhizal fungi and

Pseudomonas in reduce drought stress damage in flax (*Linum usitatissimum* L.): A field study. Mycorrhiza. 2017;**27**:537-552. DOI: 10.1007/s00572-017-0775-y

[107] Bárzana G, Aroca R, Ruiz-Lozano JM. Localized and nonlocalized effects of arbuscular mycorrhizal symbiosis on accumulation of osmolytes and aquaporins and on antioxidant systems in maize plants subjected to total or partial root drying. Plant, Cell & Environment. 2015;**38**:1613-1627. DOI: 10.1111/pce.12507

[108] Miller G, Suzuki N, Ciftci-Yilmaz S, Mittler R. Reactive oxygen species homeostasis and signalling during drought and salinity stress. Plant, Cell & Environment. 2010;**33**:453-467. DOI: 10.1111/j.1365-3040.2009.02041.x

[109] Caverzan A, Passaia G, Rosa SB, Ribeiro CW, Lazzarotto F, Margis-Pinheiro M. Plant responses to stresses: Role of ascorbate peroxidase in the antioxidant protection. Genetics and Molecular Biology. 2012;**35**:1011-1019

[110] Wu QS, Zou YN, Abd-Allah EF. Mycorrhizal association and ROS in plants. In: Ahmad P, editor. Oxidative Damage to Plants. Cambridge: Academic Press; 2014. pp. 453-475

[111] Eftekhari M, Alizadeh M, Ebrahimi P. Evaluation of the total phenolics and quercetin content of foliage in mycorrhizal grape (*Vitis vinifera* L.) varieties and effect of postharvest drying on quercetin yield. Industrial Crops and Products. 2012;**38**:160-165. DOI: 10.1016/j.indcrop.2012.01.022

[112] Zubek S, Rola K, Szewczyk A, Majewska ML, Turnau K. Enhanced concentrations of elements and secondary metabolites in *Viola tricolor* L. induced by arbuscular mycorrhizal fungi. Plant and Soil. 2015;**390**:129-142. DOI: 10.1007/s11104-015-2388-6

[113] Heidari Z, Nazari Deljou MJ. Improvement of morphophysiological traits and antioxidant capacity of zinnia (Zinnia elegance dreamland red) by arbuscular mycorrhizal fungi (*Glomus mosseae*) inoculation. International Journal of Advanced Biological and Biomedical Research. 2014;**2**:2627-2631

[114] Bai T, Li C, Ma F, Feng F, Shu H. Responses of growth and antioxidant system to root-zone hypoxia stress in two *Malus* species. Plant and Soil. 2010;**327**:95-105. DOI: 10.1007/s11104-009-0034-x

[115] Guo WL, Chen RG, Gong ZH, Yin YX, Ahmed SS, He YM. Exogenous abscisic acid increases antioxidant enzymes and related gene expression in pepper (*Capsicum annuum*) leaves subjected to chilling stress. Genetics and Molecular Research. 2012;**11**:4063-4080. DOI: 10.4238/2012.September.10.5

[116] Boaretto LF, Carvalho G, Borgo L, Creste S, Landell MG, Mazzafera P, et al. Water stress reveals differential antioxidant responses of tolerant and non-tolerant sugarcane genotypes. Plant Physiology and Biochemistry. 2014;**74**:165-175. DOI: 10.1016/j.plaphy.2013.11.016

[117] Ortiz N, Armada E, Duque E, Roldán A, Azcón R. Contribution of arbuscular mycorrhizal fungi and/or bacteria to enhancing plant drought tolerance under natural soil conditions: Effectiveness of autochthonous or allochthonous strains. Journal of Plant Physiology. 2015;**174**:87-96. DOI: 10.1016/j.jplph.2014.08.019

[118] Mayer Z, Juhász Á, Vo Trung A, Posta K. Impact of arbuscular mycorrhizal fungi on some defense enzyme activities at early stage of maize (*Zea mays* L.) under different abiotic stresses. Applied Ecology and Environmental Research. 2019;**17**(3):6241-6253. DOI: 10.15666/aeer/1703_62416253

[119] Singh PK, Singh M, Tripathi BN. Glomalin: An arbuscular mycorrhizal fungal soil protein. Protoplasma. 2012;**250**(3):663-669. DOI: 10.1007/s00709-012-0453-z

[120] Rillig MC. Arbuscular mycorrhizae, glomalin, and soil aggregation. Canadian Journal of Soil Science. 2004;**84**:355-363. DOI: 10.4141/S04-003

[121] Verbruggen E, Van Der Heijden MG, Weedon JT, Kowalchuk GA, Roeling WF. Community assembly, species richness and nestedness of arbuscular mycorrhizal fungi in agricultural soils. Molecular Ecology. 2012;**21**:2341-2353. DOI: 10.1111/j.1365-294X.2012.05534.x

[122] Nichols KA. Indirect contributions of AM fungi and soil aggregation to plant growth and protection. In: Siddiqui ZA, Akhtar MS, Futai K, editors. Mycorrhizae: Sustainable Agriculture and Forestry. Berlin: Springer Science; 2008. pp. 177-194

[123] Mardhiah U, Caruso T, Gurnell A, Rillig M. Arbuscular mycorrhizal fungal hyphae reduce soil erosion by surface water flow in a greenhouse experiment. Applied Soil Ecology. 2016;**99**:137-140. DOI: 10.1016/j.apsoil.2015.11.027

[124] Ji L, Tan W, Chen X. Arbuscular mycorrhizal mycelial networks and glomalin-related soil protein increase soil aggregation in Calcaric Regosol under well-watered and drought stress conditions. Soil and Tillage Research. 2019;**185**:1-8. DOI: 10.1016/j.still.2018.08.010

[125] Mickan B. Mechanisms for alleviation of plant water stress involving arbuscular mycorrhizas. In: Solaiman ZM, Abbott LK, Varma A, editors. Mycorrhizal Fungi: Use in Sustainable Agriculture and Land Restoration. London: Springer-Verlag Berlin Heidelberg; 2014. pp. 225-239. DOI: 10.1007/978-3-662-45370-4_14

[126] Augé RM. Water relation, drought and vesicular-arbuscular mycorrhizal symbiosis. Mycorrhiza. 2001;**11**:3-42. DOI: 10.1007/s005720100097

[127] Bitterlich M, Sandmann M, Graefe J. Arbuscular mycorrhiza alleviates restrictions to substrate water flow and delays transpiration limitation to stronger drought in tomato. Frontiers in Plant Science. 2018;**9**:154. DOI: 10.3389/fpls.2018.00154

[128] Ghorchiani M, Etesami H, Alikhani HA. Improvement of growth and yield of maize under water stress by co-inoculating an arbuscular mycorrhizal fungus and a plant growth promoting rhizobacterium together with phosphate fertilizers. Agriculture, Ecosystems and Environment. 2018;**258**:59-70. DOI: 10.1016/j.agee.2018.02.016

[129] Abdel-Salam E, Alatar A, El-Sheikh MA. Inoculation with arbuscular mycorrhizal fungi alleviates harmful effects of drought stress on damask rose. Saudi Journal of Biological Sciences. 2018;**25**:1772-1780. DOI: 10.1016/j.sjbs.2017.10.015

[130] Grümberg BC, Urcelay C, Shroeder MA, Vargas-Gil S, Luna CM. The role of inoculum identity in drought-stress mitigation by arbuscular mycorrhizal fungi in soybean. Biology and Fertility of Soils. 2015;**51**:1-10. DOI: 10.1007/s00374-014-0942-7

[131] Rahimzadeh S, Pirzad A. *Pseudomonas* and mycorrhizal fungi co-inoculation alter seed quality of flax under various water supply conditions. Industrial Crops and Products. 2019;**129**:518-524. DOI: 10.1016/j.indcrop.2018.12.038

[132] Oliveira RS, Carvalho P, Marques G, Ferreira L, Pereira S, Nunes M, et al.

Improved grain yield of cowpea (*Vigna unguiculata*) under water deficit after inoculation with *Bradyrhizobium elkanii* and *Rhizophagus irregularis*. Crop & Pasture Science. 2017;**68**:1052-1059. DOI: 10.1071/CP17087

[133] Salvioli A, Zouari I, Chalot M, Bonfante P. The arbuscular mycorrhizal status has an impact on the transcriptome profile and amino acid composition of tomato fruit. BMC Plant Biology. 2012;**12**:44. DOI: 10.1186/1471-2229-12-44

[134] Baslam M, Goicoechea N. Water deficit improved the capacity of arbuscular mycorrhizal fungi (AMF) for inducing the accumulation of antioxidant compounds in lettuce leaves. Mycorrhiza. 2012;**22**:347-359. DOI: 10.1007/s00572-011-0408-9

[135] Amiri R, Nikbakht A, Rahimmalek M, Hosseini H. Variation in the essential oil composition, antioxidant capacity, and physiological characteristics of *Pelargonium graveolens* L. inoculated with two species of mycorrhizal fungi under water deficit conditions. Journal of Plant Growth Regulation. 2017;**36**(2):502-515. DOI: 10.1007/s00344-016-9659-1

[136] Rodriguez A, Sanders IR. The role of community and population ecology in applying mycorrhizal fungi for improved food security. The ISME Journal. 2015;**9**:1053-1061. DOI: 10.1038/ismej.2014.207

[137] Faye A, Dalpé Y, Ndung'u-Magiroi K, Jefwa J, Ndoye I, Diouf M, et al. Evaluation of commercial arbuscular mycorrhizal inoculants. Canadian Journal of Plant Science. 2013;**93**:1201-1208. DOI: 10.4141/cjps2013-326

[138] Labidi S, Jeddi FB, Tisserant B, Yousfi M, Sanaa M, Dalpé Y, et al. Field application of mycorrhizal bio-inoculants affects the mineral uptake of a forage legume (*Hedysarum coronarium* L.) on a highly calcareous soil. Mycorrhiza. 2015;**25**:297-309. DOI: 10.1007/s00572-014-0609-0

[139] Briccoli Bati C, Santilli E, Lombardo L. Effect of arbuscular mycorrhizal fungi on growth and on micronutrient and macronutrient uptake and allocation in olive plantlets growing under high total Mn levels. Mycorrhiza. 2015;**25**:97-108. DOI: 10.1007/s00572-014-0589-0

[140] Estrada B, Aroca R, Barea JM, Ruiz-Lozano JM. Native arbuscular mycorrhizal fungi isolated from a saline habitat improved maize antioxidant systems and plant tolerance to salinity. Plant Science. 2013;**201-202**:42-51. DOI: 10.1016/j.plantsci.2012.11.009

[141] Silvani VA, Bidondo LF, Bompadre MJ, Colombo RP, Pérgola M, Bompadre A, et al. Growth dynamics of geographically different arbuscular mycorrhizal fungal isolates belonging to the '*Rhizophagus* clade' under monoxenic conditions. Mycologia. 2014;**106**(5):963-975. DOI: 10.3852/13-118

[142] Vosátka M, Albrechtová J. Benefits of arbuscular mycorrhizal fungi to sustainable crop production. In: Khan MS, Zaidi A, Musarrat J, editors. Microbial Strategies for Crop Improvement. Berlin Heidelberg: Springer-Verlag; 2009. pp. 205-225. DOI: 10.1007/978-3-642-01979-1_10

[143] Van Geel M, De Beenhouwer M, Lievens B, Honnay O. Crop-specific and singlespecies mycorrhizal inoculation is the best approach to improve crop growth in controlled environments. Agronomy for Sustainable Development. 2016;**36**:37. DOI: 10.1007/s13593-016-0373-y

[144] Borriello R, Lumini E, Girlanda M, Bonfante P, Bianciotto V. Effects of different management practices on arbuscular mycorrhizal fungal

diversity in maize fields by a molecular approach. Biology and Fertility of Soils. 2012;**48**:911-922. DOI: 10.1007/s00374-012-0683-4

[145] Davison J, Öpik M, Zobel M, Vasar M, Metsis M, Moora M. Communities of arbuscular mycorrhizal fungi detected in forest soil are spatially heterogeneous but do not vary throughout the growing season. PLoS One. 2012;7:e41938. DOI: 10.1371/journal.pone.0041938

[146] Hernádi I, Sasvári Z, Albrechtová J, Vosatka M, Posta K. Arbuscular mycorrhizal inoculant increases yield of spice pepper and affects the indigenous fungal community in the field. HortScience. 2012;**47**(5):603-606. DOI: 10.21273/HORTSCI.47.5.603

[147] Berruti A, Borriello R, Della Beffa MT, Scariot V, Bianciotto V. Application of non specific commercial AMF inocula results in poor mycorrhizationin *Camellia japonica* L. Symbiosis. 2013;**61**:63-76. DOI: 10.1007/s13199-013-0258-7

[148] Berruti A, Borriello R, Lumini E, Scariot V, Bianciotto V, Balestrini R. Application of laser microdissection to identify the mycorrhizal fungi that establish arbuscules inside root cells. Frontiers in Plant Science. 2013;**4**:135. DOI: 10.3389/fpls.2013.00135

[149] Thonar C, Erb A, Jansa J. Realtime PCR to quantify composition of arbuscular mycorrhizal fungal communities-marker design, verification, calibration and field validation. Molecular Ecology Resources. 2012;**12**:219-232. DOI: 10.1111/j.1755-0998.2011.03086.x

[150] Walder F, vanderHeijden MGA. Regulation of resource exchange in the arbuscular mycorrhizal symbiosis. Nature Plants. 2015;**1**:15159. DOI: 10.1038/nplants.2015.159

[151] Sawers RJH, Gutjahr C, Paszkowski U. Cereal mycorrhiza: An ancient symbiosis in modern agriculture. Trends in Plant Science. 2008;**13**:93-97. DOI: 10.1016/j.tplants.2007.11.006

Chapter 11

Physiological Features of Red Currant Adaptation to Drought and High Air Temperatures

Panfilova Olga Vitalevna, Golyaeva Olga Dmitrievna, Knyazev Sergey Dmitrievich and Kalinina Olga Vitalevna

Abstract

An important requirement for varieties is adaptation to growing conditions. The main indicators of water regime, photosynthesis, and productivity of representatives of different species of *Ribesia* (Berl.) Jancz. subgenus to drought and high temperatures have been studied. Quantitative and qualitative changes of the photosynthetic apparatus are the response to drought and high temperatures. The ratio of chlorophylls to carotenoids is considered to be one of the indicators of adaptability. The total water potential in red currant leaves depends on shoot growth, leaf age, berry formation, variety, and weather conditions, and it is not the main indication of drought resistance. The ratio of bound and free water and water-holding capacity of the leaves is considered to be a determining sign of resistance to hyperthermia. Red currant genotypes do not possess high indicators of heat resistance. The prospects of using physiological rapid diagnostic methods in breeding for adaptability to destructive factors of the growing season are shown. The representatives of *Ribes petraeum* Wulf. ("Hollandische Rote") and *R. multiflorum* Kit. (1426-21-80) have high levels of drought resistance, making them highly potential for wider growing in (semi)arid agroecological condition.

Keywords: *Ribesia* (Berl.) Jancz. subgenus, genotypes, drought resistance, heat hardiness, water regime

1. Introduction

The ecological factors of the environment play an important role in the resistance of fruit plants, their productivity, and crop quality [1, 2]. Global climate change is now threatening. Air and soil summer droughts reduce the quality and quantity of horticultural products. Weather anomalies lead to an imbalance of the protective mechanisms of fruit and berry crops and adversely affect the physiological processes occurring in plants [3, 4]. Recently, the task to obtain new high-yielding, precocious genotypes with an amplitude of adaptation to different conditions, with a high content of nutrients and biologically active substances in the fruit, is faced by scientists from different countries [5]. The search of express methods of plant diagnostics to destructive factors of the environment is a priority direction. It is known that the temperature increase causes morphological and anatomical, physiological, and biochemical changes, which affect the growth and development of plants and can lead to large economic losses

[6, 7]. The study of morphological features, structure of photosynthetic apparatus, and water exchange of plants in connection with the area of growth is the main condition for solving fundamental and applied problems in the biology of the culture. The use of physiological and biochemical rapid methods of diagnostics of plant resistance to adverse weather and climatic factors can significantly optimize the long breeding process, minimize crop losses, and obtain genotypes resistant to the destructive effects of climatic anomalies [8–10]. Plant organisms have different mechanisms of adaptation to stressors [9, 11]. Xeromorphic structure of a plant leaf, changes in a pigment complex, and water balance are important diagnostic signs of drought resistance and heat resistance [12–14]. Issues of adaptation of berry crops to drought and high temperatures are poorly studied. Red currant is one of the valuable berry crops due to the high content of vitamins, microelements, sugars, and organic acids. It is valued as a source of healthy nutrition [15–18]. Vitamin and the healing properties of the berries of this culture are also preserved in processed products [19]. Introduction is considered to be an important link in the distribution and production of new red currant genotypes. The success of the introduction is determined by the nature of the interaction of hereditary biological characteristics of plants with specific environmental conditions [18]. Red currants belong to the *Ribes* L. genus and *Ribesia* (Berl.) Jancz. subgenus. As a culture, it was developed on the basis of four species, i.e., *Ribes vulgare* Lam., *Ribes petraeum* Wulf., *Ribes multiflorum* Kit., and *Ribes rubrum* L., and their hybrids [20]. The world assortment of the *Ribes* L. genus includes more than 200 varieties; however, the genetic resources of the *Ribesia* (Berti.) Jancz. subgenus are poorly studied, since there are a number of wild species that exceed the existing varieties by a number of economic and biological characteristics [21].

2. Anatomical, morphological, and physiological changes of a red currant leaf due to drought and high temperature

The study of the assimilation apparatus in red currants gives a complete picture of the plasticity of the genotype in relation to the ecological factors of the environment [11, 22–24]. The investigation of the anatomical and morphological features of red currant leaf structure showed that some morphological features, such as shape and venation, depend on the biological characteristics of the variety and the anatomical structure is more influenced by the growing conditions. During the periods with insufficient water supply, the leaf surface area decreased in all studied currant samples compared to optimal conditions. A positive relationship between leaf area and hydrothermal coefficient ($r = +0.99$) and a negative relationship between leaf area and ambient temperature ($r = -0.97$) were identified [10]. Rezanova [14], Patzukova [25], Tokhtar [18], and Panfilova [2] studied the anatomical structure of currant leaves and described the structure of the stomatal apparatus and the conduction system of *Ribes rubrum* L. and *Ribes American* L. According to anatomical studies, the red currant leaf had a mesomorphic structure characteristic of the *Ribesia* (Berl.) Jancz. subgenus. Spongy parenchyma prevailed over palisade parenchyma, epidermal cells were large enough, and on the upper side of the leaf, they were larger than on the lower one (**Figure 1**) [10]. It is shown that stomata are formed on the lower side of the leaf rarely and unevenly, at the level of the epidermis. There are varietal species differences in the size and number of stomata per 1 mm^2 of leaf area, in the length of the closing cells, and in their shape. The size of stomata and the degree of their openness depend on the temperature and humidity of the air. During the drought, the degree of stomata opening decreases sharply.

The index of leaf mesostructure is labile in red currant genotypes of different ecological and genetic origins, including *Ribes petraeum* Wulf. ("Hollandische Rote"

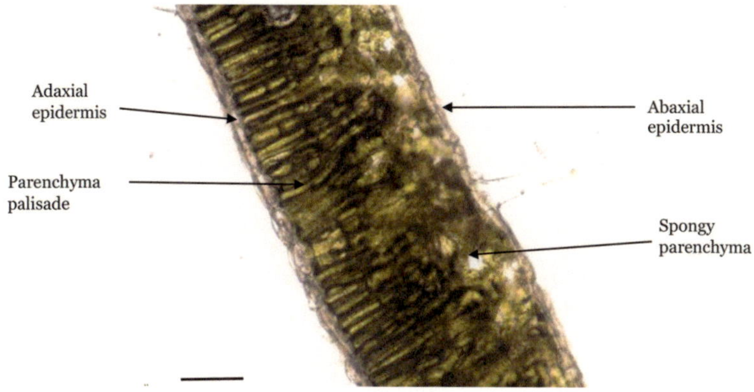

Figure 1.
Anatomical structure of a red currant leaf (cross sectional view, ×40), R. vulgare Lam. Bar represents 10 μm.

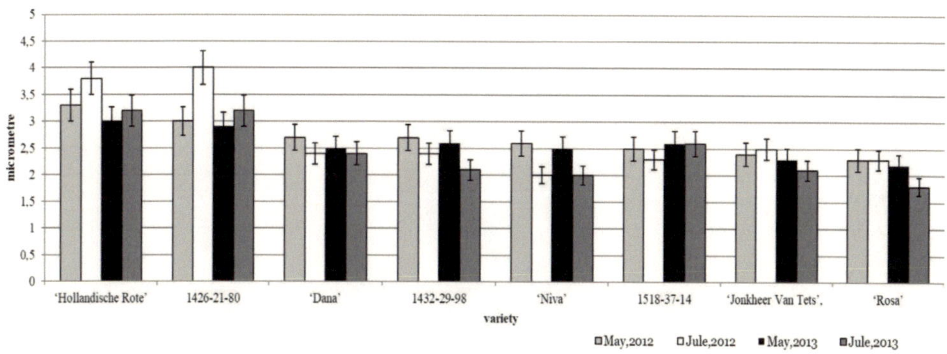

Figure 2.
The average size (±S.E.) of the cells of adaxial epidermis of a leaf in red currant varieties and selected genotypes of the Russian Research Institute of Fruit Crop Breeding, Orel [10].

and 1518-37-14), *Ribes vulgare* Lam. ("Jonkheer Van Tets," "Niva" and "Rosa"), and *Ribes multiflorum* Kit. ("Dana," 1426-21-80 and 1432-29-98), depending on the genotypes and phase of plant development.

High temperature and drought had different effects on cell size of adaxial epidermis and leaf mesostructure. In drought periods (2012, 2013) at the stages of active growth of shoots and formation of berries, the main cells of the adaxial epidermis in "Hollandische Rote" (*Ribes petraeum* Wulf) and 1426-21-80 (*Ribes multiflorum* Kit.) under the action of high temperatures (up to +31.2 ... +28.6°C in May and +32.2 ... +31.5°C in July) were somewhat stretched due to the decrease in turgor of the cells. The remaining samples showed cell compression in the tangent direction (parallel to the stem surface) (**Figure 2**).

Increased temperature and drought lead to the growth of parenchymal cells and increase of the overall thickness of the leaf (**Table 1**). The growth of mesophyll cells occurs mainly due to the increase in the volume of air-bearing cavities of the spongy parenchyma, which contributes to the improvement of gas exchange between the leaf and the environment (**Figure 3**).

In some previous studies [2, 26, 27], it was confirmed that significantly greater growth of spongy parenchyma cells and leaf blade thickness in dry periods were found in most samples of *Ribesia* (Berl.) Jancz. (with the exception of "Niva"). The largest increase in the thickness of the leaf was noted in "Hollandische Rote" and 1426-21-80. The authors consider these changes as a manifestation of high

Variety name/code (B)	Year (A)								
	2011			2012			2013		
	PP	SP	TLT	PP	SP	TLT	PP	SP	TLT
1426-21-80	10.29	13.62	26.94	12.00	20.00	37.30	9.75	15.80	29.91
"Hollandische Rote"	9.56	15.24	27.70	11.25	18.45	35.35	9.55	16.95	31.40
"Niva"	9.28	12.65	24.63	10.15	15.35	29.60	10.35	15.30	29.10
"Dana"	8.10	12.00	22.70	10.05	16.30	29.92	9.10	14.85	27.70
1432-29-98	7.90	12.35	23.25	11.65	17.55	33.65	10.15	15.10	29.10
1518-37-14	7.70	14.68	25.38	9.15	18.15	30.80	8.55	16.90	29.15
"Jonkheer Van Tets"	7.30	10.10	20.33	10.50	15.90	30.05	9.00	12.30	24.80
"Rosa"	6.80	11.23	20.53	7.95	13.85	25.60	8.10	13.80	25.10

Notes: PP—palisade parenchyma, μm; SP—sponge parenchyma, μm; TLT—total leaf thickness, μm. LSD_{05} for palisade parenchyma A = 1.57, B = 2.56, and AB = 4.42; LSD_{05} for sponge parenchyma A = 1.92, B = 3.14, and AB = 5.43; LSD_{05} for total leaf thickness A = 1.39, B = 2.27, and AB = 3.93.

Table 1.
Leaf mesostructural parameters in red currant varieties from the Russian Research Institute of Fruit Crop Breeding, Orel, in 2011–2013 vegetation period [10].

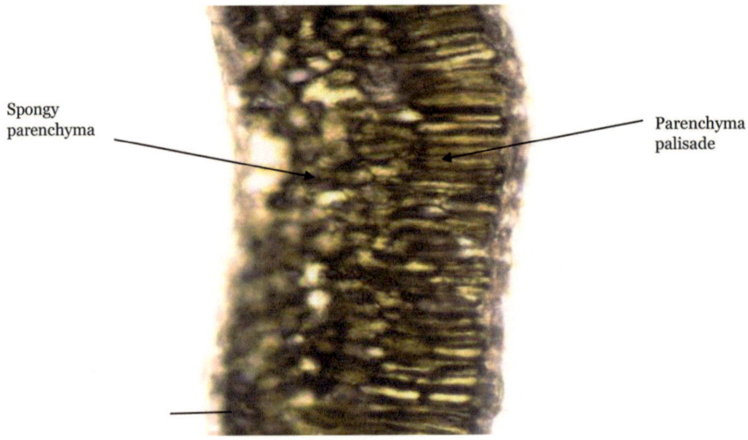

Figure 3.
The leaf mesophyll cells in a red currant exposed to drought ((cross sectional view, ×40), R. multiflorum Kit.). Bar represents 10 μm.

adaptability of the anatomical structure of the leaf to stressors (high temperature) during the growing season (**Table 1**).

Knowledge of the anatomical structure of the leaves helps to fully reveal the biological characteristics of the variety and species as well as the flow of the most important physiological processes [28]. In the process of evolution, many currant species have acquired a high potential of photosynthetic productivity, and, for the most part, in the real environment, this potential is not fully used [29, 30].

Drought is one of the reasons for the decrease in the intensity of photosynthesis, respiration, and changes in the hormonal status of plants. Violation of hormonal metabolism significantly increases the process of natural fall of the ovaries of red currants, as well as the appearance of necrotic points on the leaves and a decrease in the number of laid generative buds; leads to a decrease in yield, during both current and subsequent years; and also reduces the stability of plants in winter [31–33].

The process of photosynthesis largely depends on the quantitative and qualitative composition of pigments. Changes in the pigment complex can be considered as important mechanisms of culture adaptation to environmental conditions. The impact of high temperatures and low soil moisture leads to a decrease in the concentration of chlorophyll [22, 34–36]. Studies of different red currant species have confirmed the impact of air and soil drought on the work of the assimilation apparatus.

In the dry period (in the Orel region it was 2012), there was a decrease in the content of chlorophyll *a* and chlorophylls *a+b* in the leaves of all red currant genotypes in comparison with the favorable weather conditions (2011) (**Figure 4**).

A sharp (maximum) decrease in the number of pigments was observed in all derivatives of *R. vulgare* Lam. in 2012; the minimum decrease was observed in the representatives of *R. petraeum* Wulf. and *R. multiflorum* Kit. A slight decrease in drought in 2013 contributed to a slight increase in pigment content.

Drought at high temperatures stimulates the formation of carotenoids in the cell, which is consistent with the literature data on the protective function of this group of pigments under stress [5, 37, 38]. In the representatives of *Ribesia* (Berl.) Jancz. subgenus, the content of carotenoids in drought conditions increased by 2–3 times compared to a favorable period. A positive correlation in red currants was found between the carotenoid content and temperature (r = +0.77) [2, 10]. An indirect indicator of drought resistance and high temperature is the ratio of the sum of chlorophylls to carotenoids [23, 39, 40]. In stressful conditions, the high value of this indicator was observed in the variety "Hollandische Rote" (the value of the coefficient 5.14) and the selected form 1426-21-80 (the value of the coefficient 5.51), which may indicate the stability of these samples to drought and high summer temperatures. Hydrothermal regime affects productivity by affecting the functional state of plants, which was confirmed by high correlations between the amount of chlorophyll *a* and yield (r = +0.85) and the sum of chlorophylls and yield (r = +0.78) [10].

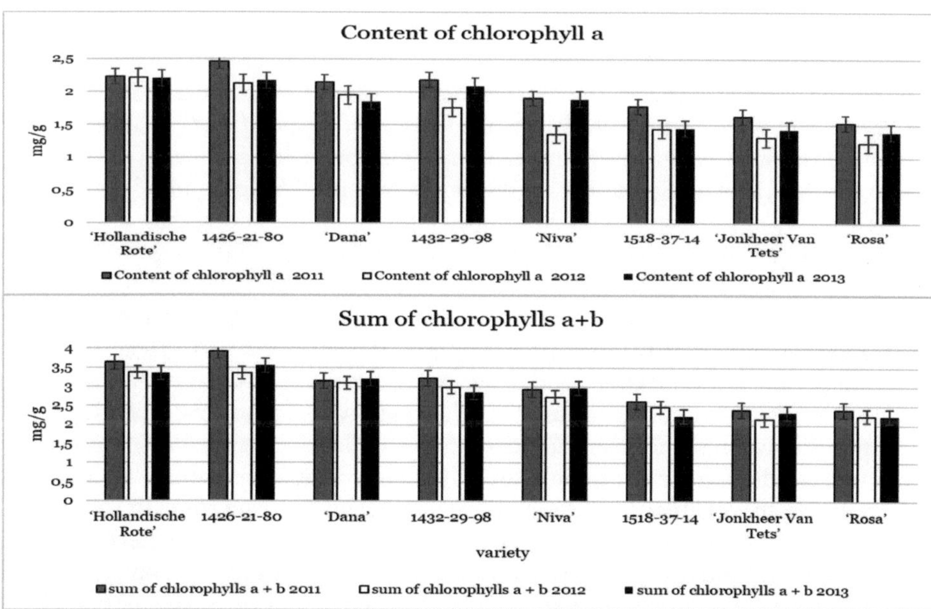

Figure 4.
Content of chlorophyll a and sum of chlorophylls a+b in red currant varieties and selected genotypes of the Russian Research Institute of Fruit Crop Breeding, Orel [10].

3. Features of water regime in red currants (drought resistance, heat hardiness)

Changes in the hydration of tissues and their water balance under the influence of adverse environmental conditions are reflected in the level, direction, and relationship of physiological processes that determine the formation of the crop and its quality. The ability to retain and economically consume water in arid conditions is a protective and adaptive reaction of resistant plants, which is due to a number of internal factors [33, 43].

It is established that the increase of water-holding capacity of leaves in extreme conditions of water supply allows stable red currant plants to regulate water metabolism [33, 41]. High water-holding capacity against the background of low water potential (61–67%) was observed in the varieties "Jonkheer Van Tets," "Hollandische Weisse," and "Englische Grosse Weisse." It is noted that the water potential of red currant leaves is not the main indicator of plant resistance to drought, because it depends on a number of factors (weather and climatic conditions, genotype, leaf age, shoot growth, and berry formation). The maximum reduction of water potential of leaf tissues in representatives of the subgenus *Ribesia* (Berl.) Jancz. occurs during the formation of berries, as most of the water is used to form the ovary [10, 42].

The resistance of red currants to hyperthermia is determined by the high content of bound water and water-holding capacity of leaves, low water deficit, and intensity of transpiration [43–45]. High values of the coefficient of bound water to free water and minimal water losses were observed in the variety "Hollandische Rote" and selected form 1426-21-80. The genotypes of these samples of *R. petraeum* Wulf. and *R. multiflorum* Kit. species exhibit higher adaptability to high temperatures against the background of soil moisture deficit (**Table 2**) [10]. In terms of transpiration intensity, red currant samples exhibit different mechanisms of adaptation to drought: either reduce the level of transpiration intensity (plants save water) or the level of transpiration intensity remains high (drought level for these genotypes is not critical) [43].

The relationship between physiological characteristics was revealed in representatives of different red currant species, which was confirmed by the coefficient of pair correlation between the amount of chlorophylls and water loss ($r = -1.00$), the amount of chlorophylls, and the amount of free water ($p = +0.98$). The content of pigments, fractional composition of water, and water-holding capacity of the leaf were associated with the development of water deficit, which depended on the air temperature ($r = +0.84$). The relationship between water regime indicators and meteorological features of the growing season was described using multiple regression coefficients (**Table 3**) [10].

In the study of plant resistance to abiotic environmental factors, their ability to withstand high air temperatures is important. Abnormally high temperatures, regardless of the dehydrating effect of dried soil and air, lead to disruption of water metabolism in plants and damage to membranes and proteins of the cell [46–50]. In this regard, an important feature of the variety is heat resistance [51]. Heat resistance largely depends on the duration of high temperatures and their absolute value. Temperature limits are specific to an individual genotype. In most cases, fruit plants begin to suffer when the temperature rises to 35–40°C. At these and higher temperatures, normal physiological functions are inhibited, and at temperatures around 50°C protoplasm, coagulation and cell death may occur. The exceeding the optimum temperature level leads to partial or global denaturation of proteins. In heat-resistant genotypes in the lipid complex, the saturated fatty acids predominate, and their appearance is a consequence of adaptation to this damaging factor. The mechanism of increasing the heat resistance is interconnected with the genetic apparatus of the cell and is aimed at stabilizing the membrane lipids in the direction of reducing their poly saturation. Under the action of elevated temperatures in the cells, the synthesis of stress proteins is induced [52]. Heat resistance is also associated with a certain

Variety name/code (B)	Year (A)							
	2011		2012		2013		Average	
	BW/FW	WL, %	BW/FW	WL, %	BW/FW	WL, %	BW/FW	WL, %
"Hollandische Rote"	1.33	35.82	2.09	20.36	2.03	21.11	1.82	25.76
1426-21-80	1.31	36.16	2.41	20.62	2.09	19.37	1.94	25.38
1432-29-98	1.12	30.17	1.42	25.01	1.20	26.98	1.25	27.39
"Jonkheer Van Tets"	1.01	38.10	1.31	21.85	1.04	27.40	1.12	29.12
1518-37-14	0.96	40.91	1.06	27.62	1.08	26.66	1.03	31.73
"Niva"	0.82	42.56	1.71	24.84	1.50	25.58	1.34	30.99
"Dana"	0.70	40.97	1.76	25.51	1.66	25.44	1.37	30.64
"Rosa"	0.63	39.75	0.98	26.82	0.87	27.85	0.83	31.47
LSD_{05}	0.08	2.06	0.11	3.03	0.14	1.86		

Notes: BW/FW—ratio of bound and free water; WL—water loss in 24 hours; % of water content, LSD_{05} for BW/FW^—A, 2.18; B, 4.59; and AB, 7.95; and LSD_{05} for WL—A, 0.13; B, 0.21; and AB, 0.37.

Table 2.
The main indicators of the water regime in red currant varieties from the Russian Research Institute of Fruit Crop Breeding, Orel, in 2011–2013 vegetation period [10].

Indicator	Regression ratio
Hydration	0.90×F; r = 0.22
Free water fraction	0.84×F; r = 0.33
Water loss	−0.88×F; r = 0.30

Notes: F—the values of hydrothermal coefficient; and r—the standard error of the experiment.

Table 3.
Correlation between physiological indicators and meteorological conditions in studied red currant varieties from the Russian Research Institute of Fruit Crop Breeding, Orel [10].

stage of plant development: young actively growing tissues are less stable than old ones. For berry plants, high temperatures are especially dangerous during flowering, because they cause sterility of flowers and the fall of ovaries. Plant organs differ in their heat resistance: shoots and buds are most stable, and the root system is less stable. The cambial tissue is most stable among tissues.

Russian and foreign researchers have achieved certain results in the study of water metabolism and physiological and biochemical parameters of resistance of fruit plants and developed methods for determining heat resistance. Water loss and the degree of water content recovery are informative indicators of heat resistance assessment [34, 53–64].

The effect of temperature "shock" +50°C on the red currant genotypes showed that this culture does not have high heat resistance. Leaves at a young age (less heat-resistant) are most severely damaged by high temperatures; the older the leaf, the higher its heat resistance [65]. The degree of water recovery in red currant leaves increases by the time of berry ripening. The leaves of red currant genotypes recover water better during berry ripening and are less susceptible to damage by temperature +50°C; they are more heat-resistant than the leaves during the active growth of shoots. One hundred percent recovery of water loss does not occur after exposure to temperature stress (+50°C) on the leaf. The most heat-resistant varieties are considered "Niva" (*R. vulgare* Lam.) and "Hollandische Rote" (*R. petraeum* Wulf.) [66].

Field assessment of red currant plants was carried out in Adygea after a long and gradually increasing heat in 1998. "Jonkheer Van Tets," "Natali," "Nenagliadnaya," "Boulogne Blanche," and *Ribes biebersteinii* Berl. showed resistance to heat (damage up to 2.5 points). The red currant samples of "Englische Grosse Weisse," "Hollandische Weisse," "Boulogne Rouge," "Versailles Rouge," "White Viksne," "Hollandische Weisse," and "London Market" dropped all the leaves, but the kidneys were still alive. Necrotic spots of different degrees appeared on the leaves of "Jonkheer Van Tets," "Nenagliadnaya," and "Boulogne Blanche" [33].

4. Conclusion

Summarizing the above, it can be approved that the problem of drought resistance and heat resistance of varieties is currently particularly acute. The increase of variety stability now gets the great relevance, since the numerous experiments prove that only the use of the adapted varieties and the transition to adaptive, biological, and low-cost technologies of cultivation of fruit-growing production are possible [67]. Many issues related to the peculiarities of the water regime and physiological and biochemical processes remain unclear. In our work we have shown that berry crops, including red currants, have complex and little-known mechanisms of adaptation to drought and abnormally high temperatures. The prospects of using physiological methods in assessing the resistance of *Ribesia* (Berl.) Jancz samples to adverse weather factors have been proved. The obtained results were fully confirmed by field observations of stability of red currant varieties and selected forms under conditions of high temperature and moisture deficit. This makes it possible to use the considered methods as a rapid diagnostics and significantly intensify the selection for adaptability. As shown by the complex physiological assessment, the derivatives of different species of the *Ribesia* (Berl.) Jancz subgenus have a different ecological plasticity, which is explained by the difference of their genetic and ecological-geographical origins. The derivatives of different species of the *Ribes petraeum* Wulf. ("Hollandische Rote") and *R. muliflorum* Kit. (1426-21-80) showed the greatest resistance to dry conditions of the vegetation period in the Central Chernozem Region that gives reason to use them in introduction. Low drought resistance was observed in the varieties and selected forms of *Ribes vulgare* Lam.

Author details

Panfilova Olga Vitalevna*, Golyaeva Olga Dmitrievna, Knyazev Sergey Dmitrievich and Kalinina Olga Vitalevna
Russian Research Institute of Fruit Crop Breeding, Orel, Russia

*Address all correspondence to: olga_280682@mail.ru

IntechOpen

© 2019 The Author(s). Licensee IntechOpen. This chapter is distributed under the terms of the Creative Commons Attribution License (http://creativecommons.org/licenses/by/3.0), which permits unrestricted use, distribution, and reproduction in any medium, provided the original work is properly cited. (cc) BY

References

[1] Gudkovsky VA. Resistance of fruit and berry plants to stress factors and ways to improve it. In: Collection of Proceedings: Ways to Improve the Sustainability of Gardening; Michurinsk. 1998. pp. 17-29

[2] Panfiliva OV. Assessment of red currant adaptability to abiotic factors of the north-west of the Central-Chernozem region [thesis]. Orel: Orel State Agrarian University; 2014

[3] Khaustovich IP. Improving the sustainability of horticulture in response to climate change in the Central-Chernozem Region [thesis]. Voronezh: Voronezh State Agrarian University named after K.D. Glinka; 2006

[4] Meriño-Gergichevich C, Reyes-Díaz M, Guerrero J, Ondrasek G. Physiological and nutritional responses in two highbush blueberry cultivars exposed to deficiency and excess of boron. Journal of Soil Science and Plant Nutrition. 2017;**17**(2):307-318

[5] Krivoruchko VP. Ecological and biological bases of increase of productivity of apple orchards of Northern Kyrgyzstan [thesis]. Bishkek: National Science Academy of the Kirghiz Republic; 1998

[6] Beck EH, Fettig S, Knake C, Hartig K, Bhattarai T. Specific and unspecific responses of plants to cold and drought stress. Journal of Biosciences. 2007;**32**:501-510. DOI: 10.1007/s12038-007-0049-5

[7] Hasanuzzaman M, Nahar K, Fujita M. Extreme temperature responses, oxidative stress and antioxidant defense in plants. In: Vahadati K, Leslie C, editors. Abiotic Stress-Plant Response and Applications in Agriculture. Rijeka, Croatia: Intech; 2013. DOI: 10.5772/54833

[8] Wise RR, Olson AJ, Schrader SM, Sharkey TD. Electron transport is the functional limitation of photosynthesis in field-grown Pima cotton plants at high temperature. Plant, Cell & Environment. 2004;**27**(6):717-724. DOI: 10.1111/j.1365-3040.2004.01171.x

[9] Neeru K, Kalpna B, Kadambot HMS, Harsh N. Food crops face rising temperatures: An overview of responses, adaptive mechanisms, and approaches to improve heat tolerance. Cogent Food & Agriculture. 2016;**2**(1):271-313. DOI: 10.1080/23311932.2015.1134380

[10] Panfilova OV, Golyaeva OD. Physiological features of red currant varieties and selected seedling adaptation to drought and high temperature. Agrobiology. 2017;**52**(5):1056-1064. DOI: 10.15389/agrobiology.2017.5.1056rus

[11] Davlatov SH, Ashurov AA, Baykova EV. Morphological and anatomical leaf structure of wild species of *Berberis* genus from Tajikistan. Tomsk State University Journal;**323**:348-350

[12] Skhalyakho TV, Nenko NI, Kisileva GK. Assessment of drought resistance degree of grape varieties in the conditions of the Anapa-Taman zone by physiological, biochemical, anatomical and morphological parameters. Fruit Growing and Viticulture of South Russia. 2012;**17**:69-78

[13] Rezanova TA, Sorokopudov VN. Some features of anatomical and morphological structure of *Ribes americanum* mill. In: Proc. Sci. Conf. Biological Active Compounds of Natural Origin: Phytotherapy, Pharmaceutical Marketing, Pharmaceutical Technology, Botany; Belgorod. 2008. pp. 133-135

[14] Rezanova TA. Morphological, anatomical and ecological features of

Ribes americanum Mill. at introduction in the South of the Central Russian upland [thesis]. Saratov: Saratov State University named after N.G. Chernyshevsky; 2010

[15] Määtä K, Kamal-Eldin A, Törrönen R. Phenolic compounds in berries of black, red, green, and white currants (Ribes sp.). Antioxidants & Redox Signaling. 2001;**3**:981-993. DOI: 10.1089/152308601317203521

[16] Pantelidis GE, Vasilakakis M, Manganaris GA, Diamantidis G. Antioxidant capacity, phenol, anthocyanin and ascorbic acid contents in raspberries, blackberries, red currants, gooseberries and cornelian cherries. Food Chemistry. 2007;**102**:777-783. DOI: 10.1016/j.foodchem.2007.06.021

[17] Skrede G, Martinsen BK, Wold AB, Birkeland SE, Aaby K. Variation in quality parameters between and within 14 Nordic tree fruit and berry species. Acta Agriculturae Scandinavica Section B Soil and Plant Science. 2011;**62**:193-208. DOI: 10.1080/09064710.2011.598543

[18] Tokhtar LA. Biological features of red currants of Ribesia (Berl.) Jancz. subgenus at introduction in the conditions of Belgorod region [thesis]. Belgorod: Belgorod State National Research University; 2011

[19] Makarkina MA. Apple and red currant breeding for improvement of fruit chemical composition [thesis]. Bryansk: Bryansk State Agricultural Academy; 2010

[20] Fedorovsky VD. *Ribes spicatum* Robson. (Taxonomy, Geography, Variability, Introduction); Kiev, 2001

[21] Golyaeva OD. Red currant breeding results at the Russian Research Institute of Fruit Crop Breeding. In: Collection of Proceedings: The Current State of Currant and Gooseberry Crops; Michurinsk. 2007. pp. 32-35

[22] Bacelar EA. Ecophysiological Responses of Olive (*Olea europaea* L.) to Restricted Water Availability: Limitations, Damages and Drought Resistance Mechanisms. Vila Real: Universidade De Trás-Os-Montes E Alto Douro; 2006

[23] Kiseleva NS. Assessment of adaptive capacity of different pear genotypes according to morphoanatomic and physiological state of leaves. Agricultural Biology. 2009;**3**:34-38

[24] Krokhmal II, Pugacheva AY. Leaf anatomical features of species of *Hemerocallis* L. genus under the conditions of steppe zone of the Ukraine. Journal Bulletin of Problems Biology and Medicine. 2011;**1**(2):72-76

[25] Patzukova NG. Biological features of *Ribes Alpinum* L. at introduction in Belgorod region [thesis]. Saratov: Saratov State University named after N.G. Chernyshevsky; 2010

[26] Panfilova OV, Golyaeva OD. Currant adaptation to drought and anomalously high temperatures (overview). Contemporary Horticulture. 2015;**2**(14):88-98

[27] Golyaeva OD, Panfilova OV. Adaptive red currant varieties of VNIISPK breeding. Pomiculture and Small Fruits Culture in Russia. 2014;**38**(1):96-100

[28] Ondrasek G, Clode PL, Kilburn MR, Guagliardo P, Romić D, Rengel Z. Zinc and cadmium mapping in the apical shoot and hypocotyl tissues of radish by high-resolution secondary ion mass spectrometry (NanoSIMS) after short-term exposure to metal contamination. International Journal of Environmental Research and Public Health. 2019;**16**:373

[29] Zhidekhuna TV. Photosynthetic potential of productivity of black

currant varieties of VNIIS breeding. In: Proc. Sci. Conf.; 22-24 April 2013. Voronezh: Voronezh State Agrarian University; 2013. pp. 65-70

[30] Zhidekhuna TV. Photosynthetic active radiation and its use by different black currant varieties. In: Proc. Sci. Conf. Contemporary Cultivars and Technologies for Intensive Orchards; 15-18 July 2013; Orel. Orel: VNIISPK; 2013. pp. 93-95

[31] Shmadlak J. Flower development in pome and stone breeds. In: Kudryavtzeva RP, editor. Physiology of Fruit Plants; Moscow. 1983. pp. 265-276

[32] Borovkov VV. Biochemical aspects of ripening and falling of black currant fruit [thesis]. Moscow: Moscow Agricultural Academy named after K.A. Tbvbryazev; 1997

[33] Semenova LG, Bzhetzeva NR. Features of Productivity of Black and Red Currants in Adygea; Maykop; 2003. 143 p

[34] Kushnirenko MD, Pecherskaya SN. Physiology of Water Exchange and Drought Resistance of Plants; Kishinev, 1991.

[35] Ananina VM, Zholkevich VN. Questions of plant resistance physiology. Agricultural Biology. 2003;**1**:120-124

[36] Belous OG. Diagnostics of resistance of tea plants to stress factors. Vestnik of the Russian Agricultural Science. 2008;**6**:41-43

[37] Kiseleva NS. Influence of environmental stress factors on the morpho-anatomical structure of pear leaf blade. In: Proc. Sci. Conf. Adaptive Potential and Quality of Products of Fruit Cultivars and Cultivar-Rootstock Combinations; 24-7 July 2012; Orel. Orel: VNIISPK; 2012. pp. 115-122

[38] Khvostov DS, Kirtbaya EK. Adaptation potential of strawberries in the south of Russia. Horticulture and Viticulture. 1999;**4**:16-17

[39] Nenko NI, Ilyina IA, Skhalyakh TV. Adaptation resistance of grapes in summer. In: Proc. Sci. Conf. High-Precision Technologies of Production, Storage and Processing of Grapes. Krasnodar: SKZNIISViV; 2010. pp. 50-59

[40] Golyaeva OD, Panfilova OV. Adaptive red currant cultivars of VNIISPK breeding. Pomiculture and Small Fruits Culture in Russia. 2014;**38**(1):96-100

[41] Bzhetzeva NR. Study of cuurant gene pool for using in production and breeding in Adygea [thesis]. Krasnodar: Russian Research Institute of Rice; 2002

[42] Golyeva OD, Petrov AV. Drought resistance of red currant varieties. In: Proc. Sci. Conf. Breeding and Variety Propagation of Orchard Crops; 19-22 June 2007; Orel. Orel: VNIISPK; 2007. pp. 64-74

[43] Panfilova OV, Golyaeva OD. Effect of drought resistance on the water regime of red currants. Pomiculture and Small Fruits Culture in Russia. 2012;**31**(2):119-126

[44] Machado S, Paulsen GM. Combined effects of drought and high temperature on water relations of wheat and sorghum. Plant and Soil. 2001;**233**(2):179-187. DOI: 10.1023/A:1010346601643

[45] Mironova LN, Denisova SG, Zeynetdinova GS, Reut AA. To the assessment of the adaptive potential of decorative perennials. In: Vestnik of Voronezh State University, Series: Geography; Geoecology. Vol. 1. 2011. pp. 157-159

[46] Saratova NA. Some regulatory mechanisms of plant adaptation to drought and high temperatures. In: Physiology of Plant Drought Resistance; Moscow. 1971. pp. 28-29

[47] Altergot VF, Mordkovich SS, Ignatiev LA. Evaluation principles of drought and heat resistance of plants. In:

Methods of Plant Resistance Evaluation to Unfavorable Conditions of the Environment; Leningrad. 1976. pp. 6-17

[48] Chentzova ES. Prospects of introduction and use of some species and clones of persimmon in the Kuban zone of fruit growing [thesis]. Krasnodar: Kuban State Agrarian University; 2008

[49] Pyanikov VT. Resistance of new cherry cultivars and hybrid seedlings to drought and high temperatures in the south of the central Chernozem zone. In: Proc. Sci. Conf. State and Prospects of Fruit Breeding and Variety Propagation; 12-15 July 2005; Orel. Orel: VNIISPK; 2005. pp. 403-407

[50] Nenko NI, Kiseleva GK, Sergeyev YI. Resistance of apple plants to abiotic stresse of winter and summer periods in the coenosis of different design. In: Proc. Sci. Conf. Biological Basis for Gardening and Vegetable Growing; 22-25 September 2010; Michurinsk. 2010. pp. 240-244

[51] Savelieva IN. Genetic and selection features and economical value of columnar apple cultivars and forms [thesis]. Ramon: Russian Research Institute of Beet and Sugar named after A.L. Mazlumov; 2012

[52] Fedulov YP, Kotlyarov VV, Dotzenko KA. Plant Resistance to Unfavorable Factors of the Environment. Krasnodar: Kuban State Agrarian University; 2015. 64 p

[53] Kushnirenko MD, Goncharova EA, Bondar EM. Methods of Studying Water Metabolism and Drought Resistance of Fruit Plants; Kishinev; 1970. 79 p

[54] Genkel PA. Physiology of heat and drought resistance of plants. Moscow; 1982. 280 p

[55] Zhuchenko AA. Plant Breeding Adaptive System (Ecological and Genetic Principles). Vol. 1. Moscow: RUDN University; 2001. 780 p

[56] Leonchenko VG, Evseyeva RP. Development of methods for predicting the resistance of fruit plants to salinization. In book of Ecological Aspects of Intensification of Agricultural Production. Vol. 1. Penza; 2002. pp. 46-48

[57] Leonchenko VG, Evseyeva RP, Zhbanova EV. Preliminary selection of promising genotypes of fruit plants for ecological and genetic resistance and biochemical value of fruits. Methodical Recommendations. Michurinsk; 2007. 72 p

[58] Saveliev NI, Savelieva NN, Yushkov AN. Promising Scab Immune Apple Cultivars. Michurinsk; 2009. 128 p

[59] Yushkov AN, Chivilev VV, Borzykh NV, Abyzov VV. Khoyzhanov A.V. Resistance of fruit and berry plants to dehydration and overheating. In: Potential and Quality of Products of Fruit Cultivars and Cultivar-Rootstock Combinations Conference on scientifically and practical; 24-27 July 2012; Orel: VNISPK; 2012. p. 287-291

[60] Panfilova OV, Bogomolova NI, Goljaeva OD, Ozherel'eva ZE. Valutazione della resistenza e bacche di calore colture nella regione centrale delle terre nere della Russia. Italian Science Review. 2013;**9**:86

[61] Panfilova OV, Ozherelieva ZE, Golyaeva OD. Comparative assessment of the adaptive potential of red currant genotypes (*Ribes rubrum* L.) to abiotic factors of the vegetation period. Plant Varieties Studying and Protection. 2014;**1**(22):24-27

[62] Matysik J, Alai Bhalu B, Mohanty P. Molecular mechanisms of quenching of reactive oxygen species by proline under stress in plants. Current Science. 2002;**82**:525-532

[63] Mitteler R. Oxidative stress, antioxidants, and stress tolerance. Trends in Plant Science. 2002;**7**:405-409. DOI: 10.1016/S1360-1385(02)02312-9

[64] Sung DY, Kaplan F, Lee KJ, Guy CL. Acquired tolerance to temperature extremes. Trends in Plant Science. 2003;**8**(3):179-187. DOI: 10.1016/S1360-1385(03)00047-5

[65] Zubov AA, Luk'yanchuk IV. Field evaluation of drought resistance of strawberry plants on wilting and drying of leaves. Agricultural Biology. 2004;**1**:116-118

[66] Panfilova OV, Ozherelieva ZE, Golyaeva OD. Assessment of resistance to extremely high temperatures of vegetation period of red currant genotypes. In: Proc. Sci. Conf; 15-18 July 2013; Orel. Orel: VNIISPK; 2013. pp. 171-172

[67] Dragavtzeva IA, Savin IY, Ovechkin SV. Resource Potential of Land in Krasnodar Region for Fruit Crop Cultivation; Krasnodar; 2005. 136 p